Sustainable Forest Management and Global Climate Change

Sustainable Forest Management and Global Climate Change

Selected Case Studies from the Americas

Edited by
Mohammed H.I. Dore

Professor of Economics, Brock University, Canada

Rubén Guevara

Director-General and CEO, CATIE, Costa Rica

Edward Elgar
Cheltenham, UK • Northampton, MA, USA

Published by
Edward Elgar Publishing Limited
Glensanda House
Montpellier Parade
Cheltenham
Glos GL50 1UA
UK

Edward Elgar Publishing, Inc.
136 West Street
Suite 202
Northampton
Massachusetts 01060
USA

A catalogue record for this book
is available from the British Library

Library of Congress Cataloguing in Publication Data
Sustainable forest management and global climate change : selected case studies from the Americas / edited by Mohammed H.I. Dore, Rubén Guevara.
Includes bibliographical references (p.).
1. Forest management—America—Case studies. 2. Sustainable forestry—America—Case studies. 3. Climatic changes—America—Case studies. I. Dore, M.H.I. II. Guevara, Rubén.

SD139 .S87 2000
333.75'097—dc21

00–034823

ISBN 1 84064 161 4

Typeset by Manton Typesetters, Louth, Lincolnshire, UK.
Printed and bound in Great Britain by MPG Books Ltd, Bodmin, Cornwall.

Contents

Preface vii
About the editors ix
List of contributors xi

Introduction 1
Mohammed Dore and Rubén Guevara

PART I BIOPHYSICAL ASPECTS OF FORESTRY AND CARBON MANAGEMENT

1 Simplified guidelines for planning sustainable and diversified forest management: a case study of the Villa Mills Demonstrative Experimental Area 19
Grace Saénz and Robin aus der Beek

2 Integrating socioeconomic and biophysical data with a GIS to plan sustainable resource management efforts in Central America 40
Steven Shultz

3 Carbon storage in soils and vegetation among forested ecosystem types in northern Ontario 63
Mark Johnston and Peter Uhlig

PART II THE ECONOMIC VALUATION OF FORESTS: BOREAL AND TROPICAL CASE STUDIES

4 The carbon cycle and the value of forests as a carbon sink: a boreal case study 79
Mohammed Dore and Mark Johnston

5 The carbon cycle and the value of forests as a carbon sink: a tropical case study 107
Octavio A. Ramirez

6 Agriculture versus forestry in northern Saskatchewan 148
Mohammed Dore, Surendra N. Kulshreshtha and Mark Johnston

7 Economic valuation of mangrove ecosystems and sub-tropical forests in Central America 166
Tania Ammour, Néstor Windevoxhel and Gustavo Sención

PART III FOREST POLICY: CASE STUDIES FROM NORTH, CENTRAL AND SOUTH AMERICA

8 The Canadian Model Forest Program: a case study of the Saskatchewan model forest 201
Surendra N. Kulshreshtha

9 Forest policy in Costa Rica 225
Olman Segura-Bonilla

10 Forest management in a semi-arid region in Brazil: a case study of the state of Rio Grande do Norte 261
Sebastião Kengen, Frans Pareyn, Newton D.E. Barcellos and Francisco C.B. Campello

Name index 271
Subject index 274

Preface

The collection of research papers included in this book was made possible by a research grant from the Inter American Institute of Global Change Research (IAI), which was set up by the US National Science Foundation, with the same standards of rigour and accountability as for any grant made by the NSF itself. The IAI is one of three research and funding networks in the world dedicated to increasing the understanding of global climate change. One of the objectives of the IAI was to bring together researchers from South, Central and North America in order to increase the understanding of global climate change, and to formulate policies to mitigate global change. As principal investigators of that research grant, we the editors were able to organize a workshop at the *Centro Agronómico Tropical de Investigación y Enseñanza* (CATIE), in Turrialba, Costa Rica, on 27 February to 1 March 1997. Early versions of chapters in this book were presented and discussed at this workshop. The workshop was also co-funded by CATIE and by Brock University, in Canada.

The year 1997 was an exciting year. Only five years earlier the Rio Summit, also dubbed the 'Earth Summit', had given new optimism to the possibilities of international cooperation in the area of the conservation of the global environment. The Montreal Protocol has been an unprecedented success and the subsequent revisions to it augured well for the possible implementation of the UN Framework Convention for Climate Change (UN FCCC). At that time, we were still looking forward to the Kyoto Protocol, which was to follow in December 1997, pursuant to the articles of the UN FCCC. There was also talk of the possibility of the international community signing a global Forest Convention, which would enshrine the principles of the sustainable management of the forests of the world, an action that would aid the objectives of the UN FCCC.

The idea of a Forest Convention did not get off the ground, but the Kyoto Protocol was signed, binding the most developed countries (the so-called 'Annex II' countries) to reducing their greenhouse gas (GHG) emissions to roughly 6 per cent below their 1990 emission levels. The Protocol has yet to be ratified by the largest industrialized countries. In October 1999, the US Congress refused to ratify the Nuclear Nonproliferation Treaty, and the US Congress is on record that it will also refuse to ratify the Kyoto Protocol when it is asked to do so, unless India, China and other developing countries also accept similar emission reduction targets. Thus, instead of possible concerted global action, the best we can hope for, at least for the time being, is piecemeal actions, by individual countries, to attempt to come to terms with global climate change. Unfortunately, the adverse impacts of global climate change are expected to be badly skewed against the *poorest nations of the world*. Consequently, there appears to be little to motivate the large industrialized countries of the northern hemisphere to take action even to implement the modest GHG reduction targets agreed upon in the Kyoto Protocol. When this Preface was being written, the fifth meeting of the Conference of the Parties of the UN FCCC (COP5) had recently taken place in Bonn (25 October–5 November 1999). It seems that no substantial results were achieved at COP5, and the key questions on the implementation of the Kyoto Protocol and the Clean Development Mechanism were deferred for a year, to COP6.

The GHG reduction targets can be nicely complemented by a sensible policy of forest protection. The greatest danger of deforestation is in the tropics, and a GHG reduction policy in the developed countries could be tied to a policy of forest conservation and its judicious management. This is where the research reflected in this book might be useful. Readers will find that the authors of the chapters of this book have amassed quite a lot of scientific expertise on the questions of forest management. We also hope that the international conservation movement as a whole and influential NGOs, such as the International Union of the Conservation of Nature (IUCN) and the World Wide Fund for Nature, will also find some use for the research highlighted in these pages.

M.H.D.
R.G.

About the editors

Mohammed Dore did his doctoral studies at Oxford University. He is Professor of Economics as Brock University and has published four books, the latest being *Global Environmental Economics*, published by Basil Blackwell in 1999. A Japanese language edition of an earlier book, *The Macrodynamics of Business Cycles*, also published by Basil Blackwell, appeared in 1996. He has published a number of articles which have appeared in *Theory and Decision*, the *Journal of Environmental Economics and Management*, *Nonlinear Dynamics, Psychology and the Life Sciences*, *Environmental Ethics*, the *Journal of Post-Keynesian Economics*, *Science and Society*, *Ambio: The Journal of the Human Environment*, *Economics Letters*, the *International Review of Applied Economics*, the *Canadian Journal of Development Studies*, *History of Political Economy*, the *Canadian Journal of Economics*, and the *Journal of Comparative Economics*. Mohammed Dore also serves on the editorial boards of several journals. He was recently appointed by UNESCO to be an editor of one of the volumes of the UNESCO *Encyclopedia of Life Support Systems*. This is a major international undertaking in response to Agenda 21, and the *Encyclopedia* has 21 Nobel Laureates on its Editorial Advisory Board.

His current research interests include climate change in Canada, and global environmental change and its distributional consequences for the rich and the poor. He has collaborated with an inter-American team of researchers from South, Central and North America on the role of forests in mitigating global warming. His research has been funded by the Social Sciences and Humanites Research Council of Canada, the National Science Foundation of the USA, the Inter American Institute of Global Change Research (which is now based in Brazil) and the National Science Foundation of Brazil. He is also a member of an international consortium of research, NEPAMA, based at the University of Brasilia.

Rubén Guevara is currently Director-General and Chief Executive Officer of the Tropical Agricultural Research and Higher Education Center (CATIE), an international institution based in Costa Rica, with the mandate to improve the well-being of humankind by applying scientific research and higher education to the development, conservation and sustainable use of natural resources in the American tropics. He received his doctoral degree from the University of Idaho and attended the Program for Management Development, an Executive Education Program of the Graduate School of Business Administration, at Harvard University.

Dr Guevara has been Chairman of the Council of the International Tropical Timber Organization (ITTO), a specialized body of the United Nations Conference on Trade and Development (UNCTAD), responsible for reconciling resource management and trade issues dealing with tropical forests. He is a member of the board of the Interamerican Economic Council, based in Washington and of the International Union of Forestry Research Organizations (IUFRO), based in Austria. He has been a member of the board – serving in various capacities – in the Latin American Energy Organization (OLADE), based in Ecuador, and the Latin American Fisheries Organization (OLDEPESCA), based in Peru, and has served in specialized instances in numerous private and international institutions or organizations dealing with agricultural research, environmental conservation, higher education and rural development, such as FAO, IUCN and the Earth Council.

He has written extensively on issues associated with forestry development, human development and economic development. He is the co-author of two books: *Research in Alliance with Nature* and *Guidelines for Sustainable Forest Industries*. His current research interests include the effect of agricultural research on poverty alleviation and environmental enhancement, as well as the valuation of environmental services *vis-à-vis* the implementation of mechanisms aimed at improving human well-being and environmental protection.

Contributors

Tania Ammour is an economist and the Olafo and Mangroves Projects Leader.

Newton D.E. Barcellos, forester, Representative of IBAMA (Brazilian Institute for Environment and Renewable Natural Resources) in the state of Rio Grande do Norte, Brazil.

Robin aus der Beek is Swiss advisor to the COSUDE Project in Nepal.

Francisco C.B. Campello, forester, Chief of National Forest of Araripe in the State of Ceará/IBAMA, Brazil.

Mark Johnston is manager of Forest Ecosystem Science, Saskatchewan Environment and Resource Management.

Sebastião Kengen, forester, Phd., formerly of IBAMA, Brazil.

Surendra N. Kulshreshtha is Professor of Agricultural Economics, University of Saskatchewan.

Frans Pareyn, forester, Chief Technical Advisor of FAO Forestry Project in Guinea-Bissau.

Octavio A. Ramirez is Assistant Professor in the Department of Agricultural and Applied Economics, Texas Tech University.

Grace Saénz is a graduate student and researcher in the Natural Forest Management Unit at CATIE.

Olman Segura-Bonilla is an economist, working with the International Center for Economic Policies for Sustainable Development (CINPE), Universidad Nacional, Costa Rica.

Gustavo Sención is a masters degree student, Biodiversity Conservation and Management Area, CATIE.

Steven Shultz is Assistant Professor of Agricultural Economics, North Dakota State University, Fargo, ND, USA.

Peter Uhlig is at the Ontario Forest Research Institute, Ontario Ministry of Natural Resources.

Néstor Windevoxhel is a member of the CATIE associated personnel, coordinator for the IUCN Wetlands and Coastal Zone Program.

Introduction

Mohammed Dore and Rubén Guevara

Within the last two decades of the 1900s, it has become clear that forests are not just timber, waiting to be cut down. Ecologists have succeeded in informing others that the forests have other vital ecological functions, such as helping regulate the hydrological cycle, the nutrient cycle and the carbon cycle. Forests also help preserve the soil from being eroded away; forests are repositories of biodiversity and, as part of the hydrological cycle, they also prevent flooding. It is now fashionable for economists to refer to these ecological functions as non-marketed 'environmental services', which they attempt to value as if they were ordinary consumable goods and services. The international community too has at last recognized the problem of global warming and the latter has become the subject of a global accord, following the signing of the United Nations Framework Convention on Climate Change and the Kyoto Protocol. The UN Framework Convention recognizes that the world's forests will have an important role to play in implementing a global strategy for reining in the global emissions of carbon, which is the main factor in global warming. It is also appreciated that conservation and growth of the forests will ease other so-called 'natural' disasters, such as floods, hurricanes, and ice storms. Indeed, one fear is that, with global warming and the possible intensification of the activity of El Niño, the severity and frequency of these disasters may be on the increase. While not all the scientific evidence is yet 100 per cent conclusive on the magnitudes of future climate change, there is a rising consensus that 'something's afoot'. As a result, the international community has now accepted the *precautionary principle*, which means that the global community should err on the side of caution. This is of course a clear step forward, but much more could have been done. For example, the global community is nowhere near to the adoption of a Global Forest Convention, although the idea for such a convention has been broached many times. Such a Forest Convention would have clear guidelines for the sustainable management of the forests of the world. In the absence of such a Convention, it is imperative to produce careful and painstaking analyses on the importance of the world's forests, and show how the forests may be harvested judiciously and in a manner that might be considered at least partially 'sustainable'. This requires that the analysis take into account the role that forests might play in mitigating global warming, in preventing soil erosion, and in helping to minimize the loss of biodiversity. We hope that this book is an example of such an analysis of a number of dimensions in the sustainable management of forests. In order to reach this objective, this book draws on the expertise in the western hemisphere on the analysis of forests.

The first part of the book concentrates on the relevant biophysical aspects of forest management, and the second on the valuation of forests for providing a carbon sink that absorbs, in part, the carbon that is released into the atmosphere through the burning of fossil fuels. The third part is a more detailed look at selected case studies on forest policy in North, Central and South America. In this Introduction, the first section provides the global setting within which the chapters of this book are set. Section two is an integrated and more detailed overview on the contents of each chapter.

SECTION 1

The total area of the world's forests is about 3.4 billion hectares; that is, one-third of the world's land surface is covered by woody vegetation (FAO, 1995). A considerable portion of the total forest area is so-called open woodland, including savannas. About half (52 per cent) of open and closed forests are in the tropics, with 30 per cent at high latitudes and 18 per cent at mid-latitudes. However, this forested area is decreasing annually by approximately 12 to 15 million hectares. Most of the annual decrease is taking place in the tropical world. In fact, in temperate and boreal regions, the forest area has slightly increased in recent decades. The main reasons for deforestation in the tropics are the conversion of forest to agricultural land and to pastures, and the cutting of fuelwood and timber.

In recent years, several international conventions or accords have addressed the problem of tropical deforestation from the various perspectives of biodiversity, climate change, ozone depletion and desertification, but, as stated before, a comprehensive accord on sustainable forestry still eludes the international community. In the circumstances, the one convention that could do something to aid forest conservation is the United Nations Framework Convention on Climate Change (UN FCCC). Its ultimate objective is the stabilization of greenhouse gas emissions in the atmosphere at a level preventing dangerous anthropogenic interference with climate, and minimizing risks to and adverse effects on ecosystems. This requires close attention to forestry. Consequently, the authors of this book review and discuss options for carbon management in the tropics in forestry and agroforestry systems, and attempt to assess the feasibility of such actions. They also discuss some potential policy implications for tropical countries.

The Intergovernmental Panel on Climate Change (IPCC) has estimated that anthropogenic emissions of CO_2 in 1995 were 5.5 ± 0.5 GtC (GtC = gigatons, or billion tons of carbon) due to fossil fuel burning and cement production, plus 1.6 ± 1.0 GtC due to deforestation. During the last decade of the 20th century, about 45 per cent of anthropogenic emission stayed in the atmosphere, the remainder being taken up by the oceans and by land ecosystems (Table I.1). It might be useful to note that, before 1750, atmospheric CO_2 was 280 parts per million by volume (ppmv); in 1999, it is 365 ppmv. It is estimated that it is increasing by 1.5 ppmv per year. If we burn up all the world's fossil fuel of 5000 GtC, the atmospheric concentration of CO_2 could increase by a factor of 10, to 2800 ppmv, even after allowing for some absorption by terrestrial sinks. If that happens, it is estimated that the mean surface temperature of the earth could rise by anywhere between 4.5 and 15 degrees Celsius. Thus, in the larger scheme of things, the role of deforestation in total annual anthropogenic emission is small, but at between 22 and 50 per cent of emissions it is not negligible either.

Parties to the UN FCCC have agreed upon and committed themselves to certain actions in order to achieve the global objective of stabilizing atmospheric greenhouse gas concentrations. In principle, this target can be achieved by (a) reducing emissions of GHGs in the atmosphere (reduction of sources) and (b) removing GHGs from the atmosphere (enhancement of sinks). Research on climate change and land use in the tropics has mainly focused on the analysis of the impact of deforestation and other land use changes on greenhouse gas emissions and carbon budgets. However, some studies have attempted to quantify the potential of sustainable land use to sequester and accumulate carbon dioxide in tropical ecosystems, and others have attempted to put dollar values on boreal forests. Both types of study are represented in this book.

It should be noted that forests could also be a source of greenhouse gases, since forest ecosystems are estimated to contain more than 80 per cent of global above-ground carbon. Indeed, the scientific community has largely agreed that the tropical forest ecosystems have been net carbon sources, while the northern temperate and boreal forests are a major carbon sink (see, for example, Kauppi *et al.*, 1992; Sedjo, 1992). Recent results, however, show that mature tropical forests can accumulate carbon at a rate of 1–2Mg/ha/yr (Lugo and Brown, 1992; Grace *et al.*, 1995), which can offset emissions from deforestation and degradation (Brown *et al.*, 1996).

The carbon pool of any terrestrial ecosystem has two main components: the total area of the ecosystem and the carbon density per unit area. Thus, in order to enhance sinks and to maintain carbon pools, we can develop actions

Table I.1 Human perturbations to the carbon budget

CO_2 sources	Flux (GtC/yr)
Fossil-fuel burning	5.5
Tropical deforestation	1.6
Total anthropogenic emissions	7.1
CO_2 sinks	
Storage in the atmosphere	3.3
Uptake by the oceans	2.0
North. hemisphere forest regrowth	0.5
CO_2 and N fertilization, climate	1.3
Total	7.1

Source: IPCC (1995).

to increase the area of these ecosystems, increase their carbon density, or both of these simultaneously. Several studies have attempted to estimate the potential land area available for afforestation, reforestation, agroforestry, conservation or other management options for carbon sequestration (for example, Houghton *et al.*, 1991; Nilsson and Schophauser, 1995). However, there are many constraints on how this can be achieved. The real availability of land for these programmes depends on economic, social, cultural and institutional factors which influence land use (Trexler and Haugen, 1995; Godoy *et al.*, 1998) (Table I.2).

The sustainable management of carbon pools in existing forests, agroforestry systems and agricultural lands offers an interesting option. The above-ground carbon content of tropical forest ecosystems varies between 25 and 250 t ha^{-1}. Thus reducing deforestation and increasing measures for forest protection offers a cost-effective way of reducing CO_2 emissions (Table I.3).

Table I.2 Land available for carbon sequestration in the tropics

Option	Area (10^6 ha)
Plantation of degraded lands	0–579
Secondary forest suitable for protection	15–858
Areas with potential for agroforestry	356–500

Source: Houghton *et al.* (1991).

Other carbon management options related to existing production systems include the incorporation of trees into agricultural systems, whether as agroforestry or as a silvopastoral system. The above-ground carbon content of agroforestry and silvopastoral systems varies between 10 and 70t/ha, and the annual carbon flux to the system is between 1 and 10 ha/yr. When the carbon stored in the soils of these systems is considered, these figures can be multiplied by a factor of two (Tables I.4 and I.5).

Table I.3 Above-ground carbon storage in tropical forest carbon storage (tons per hectare)

Forest type	Primary	Secondary
Moist forests	230	190
Seasonal forests	140	120
Dry forests	60	25

Source: Brown and Lugo (1982); Brown *et al.* (1989).

It is of primary importance that the forest ecosystems of the world are in a state where their ability to act as sinks of greenhouse gases is maintained and enhanced. The measures required to achieve this may be summarized into four main types of general action:

1. Measures against desertification, deforestation and forest destruction. These should be aimed at the stabilization of forest area, where appropriate, and, where possible, increasing it further.
2. Promotion of overall health of ecosystems. This includes especially actions against harmful effects caused by pollutants.
3. Measures against the degradation of ecosystems (unsustainable management and so on) and measures that increase the potential of forests to act as sinks of GHGs (stocking densities, amount of biomass, and so on).
4. Promotion of scientific research, monitoring and information exchange on (a) forests as sources and sinks of GHGs, and (b) sustainable management of forests.

Of course, it is obvious that forestry, agroforestry and ecosystem management in the tropics cannot alone solve the problem of global warming, as the bulk of the emissions is due to fossil fuel burning. However, together with other measures, carbon management in tropical ecosystems can be one player in the global act, even though its contribution is difficult to quantify. It should also be noted that the management of tropical ecosystems might have functions and

Table I.4 Above-ground carbon storage in agroforestry and forestry in Central America

System Tree species	Life zone	Trees N/ha	Age (years)	C Storage tC/ha	C Flux tC/ha/yr
Cliricidia sepium	TPW	330	30	51.6	1.7
Cordia alliodora	TPW	278	10	24.9	2.5
Mimosa scarabella	TPW	650	2	14.2	7.1
Fuelwood Planations					
L. leucocephala	TLD	3 800	5	28.9	5.8
Eucalyptus saligna	TPW	1 378	2.5	27.0	10.8
Trees in corrals					
Alnus acuminata	TLM	35	30	25.0	0.8

Note: TPW = tropical premontane wet forest; TLD = tropical lowland dry forest; TLM = tropical lowland moist forest.

Sources: Salazar (1984, 1986); Fassbender *et al.* (1991); Picado (1986); Martinez (1989); Canet (1986).

goals other than that of solving the problem of global warming. In addition, for very many people in developing countries, trees are the main source of energy. Forestland is also cleared for agriculture, for the production of food for a constantly increasing population. This makes forest conservation extremely difficult. These social imperatives create new challenges to the sustainable management of forests and to the maintenance of biodiversity. This is not to imply, however, that nothing can be done. Our minimum target must be the development of tools that allow a combination of these multiple objectives.

Table I.5 Carbon in Tropical Soils

Land use	Carbon storage (tC/ha)
Tropical forest	60–115
Agriculture (maize)	
Initial	35
After 50 years	9
Agroforestry	
Initial	8.9
After 9 years	24.1

Sources: Brown and Lugo (1982); Young (1989); Fassbender *et al.* (1991).

For carbon sequestration, the best carbon management option is reducing deforestation and promoting reforestation and afforestation. But there is also the question of carbon density and also the cost per ton of carbon. Reducing deforestation is also the lowest-cost means of increasing carbon sequestration. These carbon management options are summarized schematically in Table I.6.

The authors of the chapters of this book all argue that the world can do much to reduce the problem of carbon accumulation in the atmosphere by paying much greater attention to the conservation of the world's forests. As the western hemisphere still has vast tracks of forests, it is imperative that effort be devoted to a proper study of these forests. Section 2 below summarizes how each author in this book tackles the task.

Table I.6 Carbon management options and trade-offs

Option	Carbon density	Sequestration (short-term)	Cost per ton of carbon
Reducing deforestation and protecting forests	High	Low	Low
Reforestation and afforestation	Moderate	High	Moderate
Silviculture	High	Moderate	Low
Agroforestry	Low	Moderate	Moderate
Fuelwood plantations	Moderate	High	High
Forest products	Low	Low	Low

SECTION 2

The first chapter by Grace Saénz and her co-author offers an analysis of the ecological functions of forests and their respective implications for tropical forest management. They show that forest management does not merely mean ensuring the sustainable production of timber; it also means taking a series of measures to ensure the fulfilment of other ecological functions of the forests. Saénz and Robin aus der Beek concentrate on the following ecological functions: conservation of biodiversity, soil protection and the maintenance of the natural watershed. A properly conserved forest will also generate employment income, recreation and tourism benefits, and of course also provide timber. The authors draw important lessons in sustainable forest management from a case study of silviculture plots in the Villa Mills experimental area in Costa Rica. They demonstrate the importance of a specific analysis for each possible ecological function of forest management. Depending on the ecological, social, economic and cultural characteristics, forest functions will vary completely. This is another reason why a detailed analysis should be made for each specific case. The chapter points to some important lessons on the sustainable management of tropical forests.

In Chapter 2, Steven Shultz reports on two case studies that demonstrate the benefits of using GIS (geographic information systems) technology to integrate biophysical and socioeconomic data in order to plan various sustainable resource management activities. In the first case study, GIS is used to identify critical land use areas with information concerning the preferences and desires of local farmers in order to determine the benefits and costs of a reforestation programme. In the second case study, GIS is used to identify critical land uses, which were linked to socioeconomic data through the mapping of individual farm boundaries within the study area. This integration of data is then used for resource planning activities through the classification of different activities, such as soil conservation, reforestation or natural forest management, that were deemed most appropriate to the specific needs of farms, based on their size and characteristics. It also permitted the ranking of farms in order of priority for those most in need of soil conservation, based on their relative levels of land degradation.

These case studies show that natural resource professionals can use relatively simple and available methodologies to plan comprehensively the sustainable management of resources in other developing areas of the world. The major advantages of using GIS technology are the efficient identification and prioritization of critical land use areas within a watershed or study area, while taking into account both biophysical and socioeconomic data. Such integration is essential for planning and evaluating specific sustainable resource management plans over space and time. However, in order to use GIS and GPS (global positioning system) technology, national and local governments should spatially reference and maintain land use surveys, and make this information compatible with national GIS databases.

In the last chapter of Part I, Mark Johnston and Peter Uhlig present estimates of carbon storage in soils and vegetation in the forests of northern Ontario. North American boreal forests have been identified as a globally important carbon store, and are expected to play an important role in future climate change. However, landscape-scale estimates of boreal carbon storage were not available and so the authors fill in this gap by estimating carbon storage in soil and vegetation for a series of upland forested ecosystem types in northwestern Ontario, Canada, using data collected from the forest ecosystem classification system (FEC). Their estimates are for mineral soil to the depth of 100cm of forest floor, understorey and tree vegetative components. They find that the mean total carbon (C) density among all ecosystem types was 26.3 ± 8.6 kg m^{-2}, and ranged from 14.1 ± 4.5 kg m^{-2}

to 37.8 ± 13.6 kg m^{-2}. Vegetative C density was highest in old-growth red and white pine types (18.0 ± 0.9 kg m^{-2}), while soil C density was highest on wet sites dominated by deep organic deposits (26.3 ± 13.6 kg m^{-2}). In general, vegetative C density was highest on productive sites, especially those with mixed species composition. Mineral soil C was also higher on productive sites, while forest floor C was greatest on sites with low rates of decomposition, due either to a lack or an excess of moisture. A comparison of C storage estimates derived from FEC data with those from forest resource inventory data for the study indicated that FEC data yielded higher estimates due to a bias in sampling that favoured mature, fully-stocked stands. These quantitative estimates will be helpful in obtaining a more complete picture of carbon sequestration in the forest ecosystems of the northern hemisphere.

Part II deals with aspects of the valuation of forests for carbon sequestration and the valuation of other ecological services provided by forests, such as prevention of soil erosion and facilitating the nutrient cycle. The first two chapters in this part estimate the value of forests for carbon sequestration in North America and in tropical America. However, their approaches are quite different. In Chapter 4, Mohammed Dore and Mark Johnston attempt to place a value on carbon sequestration of the boreal forests of Canada. The central objective of their work is to find and estimate econometrically a shadow price or the scarcity rent of forests for mitigating global warming. Furthermore, they show that the scarcity rent must be a cardinal measure which rises with increasing scarcity of the forests. They specify an optimal control model to derive a shadow price of forests, and the value of forests is computed via econometric techniques. They fit statistical models to carbon emissions in Canada and the amount of carbon sequestered by the boreal forests of Canada in order to obtain the scarcity rent of forests for carbon sequestration. The statistical models are used to compute the marginal social opportunity cost of forests per hectare. The authors argue that it is this opportunity cost that must enter any harvesting decision model or any other social cost–benefit calculation when making valuation decisions about the forests. They estimate that this social opportunity cost of forest is a premium of about 25 per cent and the value of the forest in 1986 Canadian dollars is between $289 and $412 per hectare. This is the valuation for only one ecological service provided by forests, which is carbon sequestration.

In Chapter 5, Octavio Ramirez treats the problem in greater detail. First he distinguishes between carbon *storage*, carbon *parking* and carbon *sequestration*. Storage is holding a stock of carbon in place, as in a mature forest; parking is a temporary store, whereas sequestration is the net growth in carbon stored. He looks at three kinds of forests in the tropics (primary forests, secondary forests and plantation forests) and he assesses each as repositories of carbon. He argues that humid tropical secondary forests have a considerable potential for carbon sequestration and storage. Most of the carbon sequestration takes place during the first 20 years of growth. The per hectare economic value of the carbon sink service provided exceeds the market value of the land in most tropical humid regions of the world. Therefore, if the market value reflects the opportunity cost of the land, forest conservation is an economically sensible alternative. At a critical phase of global warming of the planet, carbon sequestration is valued and, sometimes, bilateral agreements are entered into in order to achieve the objectives of carbon sequestration. Such agreements are called *joint implementation agreements*. If a global carbon market eventually develops, promoting secondary forest regeneration and conservation can yield substantial economic benefits for developing countries, as they are rewarded for maintaining a carbon sink.

Ramirez relies on data from three experimental silviculture models: conventional harvesting (called 'control'), conventional harvesting with a posterior upper canopy treatment (called 'upper canopy'), and conventional harvesting followed by a liberation treatment, to promote the establishment of tree species with high commercial value (called 'liberation'). He finds that the upper canopy of the forest stores more carbon than the control during the first two production cycles, but posterior upper canopy treatment begins to surpass the control during the third cycle. This does not cause a significant difference in the economic value of this environmental service in the long term, because of the detrimental effect of the discount rate on the value of money in the distant future. In contrast, the much lower long-term levels of carbon storage in the liberation treatment, that remain after 60 years, have an economic cost (reduced payment) of around US$250/ha in relation to the control or upper canopy treatment.

The liberation treatment, however, shows a financial advantage in both the short and the long term, mostly due to its greater production of high-value wood during the first harvest cycle. Therefore, if forest owners get no payment for the carbon storage service, they will prefer the liberation treatment. Such a decision results in a net economic loss to society of between US$175 and US$300/ha, depending on the discount rate assumed.

This translates into several hundred million dollars when considering the extension of natural forests areas even in a small country like Costa Rica. From the standpoint of the owners, conventional harvesting would be the preferred alternative for the management of natural forests. None of the silvicultural treatments investigated are justified from the social viewpoint: liberation is handicapped by its negative effect on biomass and upper canopy for its low impact on the commercial value of the harvests.

Costa Rica is currently paying forest owners US$360/ha for the sustainable management of natural forests. This is little, even when compared just with the economic value of the environmental service of carbon storage. According to this analysis, the Costa Rican payment is not sufficient to motivate long-term sustainable management of its primary forests. The advanced compensation mechanism adopted by Costa Rica, which provides 100 per cent of the payments for carbon storage during the first five years of a contract that has a 20-year duration, is also inadequate. Otherwise, the most attractive alternative is to engage in a forest management contract, make a technically appropriate initial harvest and sell or convert the land to another use five years later, once all contract payments have been received.

Ramirez concludes that, in both the short and the long term, the sustainable management of natural primary forests is more profitable than selling or using the land for cattle or agriculture, if owners are paid near the international rate of US$10/ton for the carbon storage service provided by their forests. The compensation mechanism would then become even more critical. If credibility can be established among forest owners with respect to the fulfilment of the payment agreements, it would be preferable to distribute the payments evenly throughout the 20-year forest production cycles, making the necessary annual adjustments to compensate for inflation, so that real payments remain constant.

Overall, the analyses indicate that, if carbon sequestration is one of the main objectives of a system of forest incentives, promoting secondary forest regeneration should be a priority over primary forest conservation. This is not the case in Costa Rica's and most other developing country's incentive programmes. The fact that secondary forest ecosystems in their earlier stages of development are very vulnerable to land-use change provides an additional argument for higher incentives.

In Chapter 6, Dore, Kulshreshtha and Johnston consider the possibility of reforestation of tracts of land in the Boreal Plain Ecozone in Canada which were cleared for agricultural pursuits early in the 20th century in response to short-term high prices of grain on the international markets. They argue that the evidence on the volatility of grain prices and the soil characteristics of the land suggest that producing grain on these marginal lands is uneconomic and inefficient, and is viable only under massive subsidies from the governments. They demonstrate that the social gain from returning these lands to forests (either as agroforestry plantations or as commercial forest) exceeds the private gain of the use of these lands in producing low-yield grain and forage. This is based on comparing net social returns from each land use. The net social returns are estimated by using sophisticated econometric methods, and they avoid the use of discounting by using stochastic dominance. Having reached their conclusion that the marginal lands should be reforested, they then consider what sort of reforestation policy would enhance carbon sequestration on these lands.

In the final chapter of Part II, Tania Ammour and her co-authors consider the economic evaluation of two tropical ecosystems: the mangrove forests of the Pacific coast in Nicaragua and a subtropical forest in Guatemala. The economic value of goods as a stock corresponds to the potential gross income generated from the existing harvestable biomass in the forest, which means that natural capital is evaluated in terms of its productive capacity. The flow value of economic goods is measured according to the extraction rate of the resources, the associated costs, and the economic price of the products.

The authors begin with three hypotheses: (1) the sustainable management of these forests provides a higher value than traditional extractive practices; (2) the sustainable management of resources leads to greater value of the ecological services; and (3) the social value of the ecological functions is greater than the value of extracted commercial goods. All three hypotheses were verified in the two mangrove case studies.

The authors consider two alternative ways of managing the mangroves in Nicaragua. In the 8700ha considered, net benefit of traditional management of natural resources is US$5.7 million over 10 years, using a discount rate of 10 per cent. In the second sustainable method, net benefits are almost twice that in the traditional management (US$11 million). This demonstrates the higher value provided by ecosystems managed in a more sustainable way. The ecosystem value, on a per hectare basis, generated by goods and services is US$700 in the traditional alternative and US$1300 in the sustainable method.

In the traditional method, the large benefits obtained from fuelwood (US$ 3 249 967, or about 90 per cent of value generated by all the goods) are mainly due to the low opportunity cost of labour, implying very low production costs. As a result, goods have a higher value (US$3 596 785) when compared with services (US$2 143 365). In the sustainable method, the goods have a value almost equal to the services value. This is due to the fact that all the goods were valued, including an additional good (bark for tannin). This result confirms the importance of ecological services generated by ecosystems in the Pacific wetlands.

The chapter then reports on the authors' second case study: the economic valuation of the forest in Petén, in Guatemala. In this case study, the authors find that the stock value of the forest is US$5 072 850, equivalent to US$787/ha, while annual flow is US$195 543, equivalent to US$30/ha. With both stock and flow values, environmental services contribute the greater part of total value (80 per cent and 62 per cent respectively). The erosion control contributes 73 per cent of total stock value of environmental services, while nutrient cycling is predominant (62 per cent) when evaluating the annual forest flow values.

When considering environmental services, the social value of erosion control amounts to 59 per cent of total stock value, while carbon sequestration contributes 11 per cent. The importance of the social value of erosion control is probably due to poor soil characteristics and the valuation method used. Of the total annual flow values generated by environmental services, the social value of the nutrient cycle (as an ecological function) and, to a lesser extent, the bioversity attribute contribute 62 per cent and 39 per cent respectively. The importance of the nutrient cycle service reflects the low soil fertility and the role of fallen leaves as a main source of nutrients.

The final part of the book has three chapters. In the first, Kulshreshtha examines the Canadian Model Forest Program by focusing on the Prince Albert Model Forest (PAMF) in Saskatchewan. PAMF is a partnership of seven stakeholders, representing various levels of the government, industry and the First Nations people. The area under the jurisdiction of PAMF is small, only 315 326 hectares. In spite of the small size, the forest provides both timber and non-timber products. The goals of the model forest programme were to ensure sustainable and predictable supply of forest-based benefits (ecological, social and economic) through the management of forest ecosystems and people, and to raise awareness of the concept of sustainability, integrated resource management and ecosystem management among forest users, researchers, and managers of the forests.

The success of the model forest concept can be viewed in terms of the continuation of a working partnership among the seven stakeholders. The very fact that the original seven members decided to continue with the partnership, and added three new members in Phase 2 of the project, is enough evidence to support such a claim. In addition, during this phase, several cooperating activities were initiated. The model forest acts as a forum for bringing forth concerns of other stakeholders. However, it cannot be said with any degree of certainty whether the philosophy of forest management would change if the entire Forestry Management License Area (or the entire northern part of the province) were to be managed on these lines. Sustainable and integrated forest management also requires good information. Although significant strides have been made in collecting the missing information, many gaps persist which prevent the application of scientific knowledge that would help sustainable forest management in the region.

In Chapter 9, Olman Segura considers forest policy in Costa Rica. One of his main findings is that in the past government policies have supported mainly agriculture, cattle ranching, non-traditional products and, lately, tourism, and that the sustainable development of forestry has not been a priority. Furthermore, he finds there is very little integration between forest and industry, and even in industry there are very few examples of integration. The development of forest activities has not taken into account the intellectual, technical, cultural and economic contribution of indigenous communities and women, which could strengthen it and make it competitive.

The application of forest incentives, in spite of their importance in the development of the reforestation of the country, has not taken advantage of the opportunity to motivate the rural population to take a more active role in forest management. There are no policies to render technical assistance for the exploitation, storage and marketing of timber and non-timber products. Finally, environmental services have positive externalities for any forest, including the Costa Rican forests. In the Costa Rican case, there appears to be little recognition of these services. The predominant view of the use of natural resources is biased towards agricultural production.

In the final chapter, Sebastiao Kengen and his co-authors consider the socioeconomic importance of forest management in a semi-arid region in Brazil. They argue that the semi-arid forests of the northwest of Brazil are just as fragile as the tropical Amazon, and therefore deserve attention. The main tree found there is the 'caatinga', which, if harvested in a sustainable manner, could be valuable both as a source of timber and as a source of firewood. Its conservation would also help conserve the biodiversity of the region. While its management is still in its initial phase and still requires more research, it offers a viable alternative to forest plantations. Recognizing its potential, the 'Banco do Nordeste' has started to offer a line of credit for farmers and companies to promote forest management of the 'caatinga'. This is a great advance when one considers that traditionally the emphasis has been on forest plantations, most of which fail due to the rough climatic conditions of the northeast.

In summary, this book attempts to throw light on the various dimensions of sustainable management of forestry in North, Central and South America. The authors also argue that the economic and social value of the ecological services provided by the forests must be taken into account, as a forest is more than just timber. The authors represented in this book draw on a large number of case studies in the Americas, ranging from the boreal forest of North America and the forests of the humid tropics to the forests in the arid regions of South America.

REFERENCES

Brown, S. and A. Lugo (1982), 'Aboveground biomass estimates for tropical moist forests of the Brazilian Amazon', *Interciencia*, 17, 8–18.

Brown, S., A. Gillespie and A. Lugo (1989), 'Biomass estimation methods for tropical forests with applications to forest inventory data', *Forest Science*, 35 (4), 881–902.

Brown, S., J. Sathaye, M. Cannel and P. Kauppi (1996), 'Mitigation of carbon emissions to the atmosphere by forest management', *Commonwealth Forestry Review*, 75 (1), 80–91.

Canet, B. (1986), 'Caracteristicas del sistema silvopastoral jaul (Almus acuminata) con lecheria de altura en Costa Rica', in R. Salazar (ed.), *Simposio sobre técnicas de producción de leña en fincas pequeñas y recuperación de sitios degradados por medio de silvicultura intensiva*, Turrialba, Costa Rica, 24–8 June 1985, pp. 241–9.

FAO (1995), 'Forest resources assessment 1990: Global synthesis', *FAO Forestry Paper*, 124, Rome.

Fassbender, H., J. Beer, J. Heuveldop, A. Imbach, G. Enriquez and A. Bonneman (1991), 'Ten year balances of organic matter and nutrients in agroforestry systems at CATIE, Costa Rica', *Forest Ecology and Management*, 45, 173–83.

Godoy, J., M. Kanninen and O. Ramirez (1998), 'Análisis de los Programas de Incentivos a la Reforestación Implementados en Costa Rica', *Ponencias del Primer Congreso Latinoamericano de IUFRO*, Valdivia, Chile.

Grace, J., J. Lloyd, J. McIntyre, A. Miranda, P. Meir, H. Miranda, C. Nobre, J. Moncrieff, J. Massheder, Y. Malhi, I. Wright and J. Gash (1995), 'Carbon dioxide uptake by an undisturbed tropical rain forest in southwest Amazonia, 1992 to 1993', *Science*, 270, 778–80.

Houghton, R., D. Skole and D. Lefkowitz (1991), 'Changes in landscape of Latin America between 1850 and 1985, 2: Net release of CO_2 to the atmosphere', *Forest Ecology and Management*, 38, 173–99.

IPPC (1995), *Climate Change 1995. The Science of Climate Change*, ed. J. Houghton, L. Meira Filho, B. Callander, N. Harris, A. Kattenberg and K. Maskell, Cambridge: Cambridge University Press.

Kauppi, P., K. Mielikänen and K. Kuusela (1992), 'Biomass and Carbon Budget of European Forests, 1971 to 1990', *Science*, 256, 70–74.

Kuersten, E. and P. Burschel (1993), 'CO_2 Mitigation by Agroforestry', *Water, Air and Soil Pollution*, 70, 533–44.

Lugo, A. and S. Brown (1992), 'Tropical Forests as Sinks of Atmospheric Carbon', *Forest Ecology and Management*, 48, 83–90.

Martinez, H. (1989), 'El componente forestal en los sistemas de finca de pequeños agricultores', CATIE, Série Técnica, Boletin Técnico No. 19.

Nilsson, S. and W. Schophauser (1995), 'The Carbon-sequestration Potential of a Global Afforestation Program', *Climatic Change*, 30, 257–67.

Picado, V. (1986), '*Mimosa scrabella*, especie con potencial para sombra y producción de leña en cafetales de Costa Rica', in R. Salazar (ed.), *Simposio sobre técnicas de producción de leña en fincas pequeñas y recuperación de sitios degradados por medio de sivicultura intensiva*, Turrialba, Costa Rica, 24–8 June 1985, pp. 227–86.

Salazar, R. (1984), 'Producción de leña en árboles de Gliricidia sepium usado como sombra en cafetales de Costa Rica', *Silvoenergía* , no.2 (CATIE).

Salazar, R. (ed.) (1986), *Simposio sobre técnicas de producción de leña en fincas pequeñas y recuperación de sitios degradados por medio de silvicultura intensiva*, Turrialba, Costa Rica, 24–8 June 1985.

Sedjo, R. (1992), 'Temperate Forest Ecosystems in the Global Carbon Cycle', *Ambio*, 24 (4), 274–7.

Trexler, M. and C. Haugen (1995), *Keeping it Green: Evaluating Tropical Forestry Strategies to Mitigate Global Warming*, Washington, DC: World Resources Institute.

UN-ECE/FAO (1992), *The Forest Resources of the Temperate Zones. Main Findings of the UN-ECE/FAO 1990 Forest Resource Assessment*, ECE/TIM-60, UN-FAO, New York.

Young, A. (1989), 'Agroforestry for Soil Conservation', *ICRAF Science and Practice in Agroforestry*, 4, CAB International, Wallingford.

PART I

Biophysical Aspects of Forestry and Carbon Management

1. Simplified guidelines for planning sustainable and diversified forest management: a case study of the Villa Mills Demonstrative Experimental Area

Grace Saénz and Robin aus der Beek

INTRODUCTION

The uncertainty which exists regarding long-term forest development, along with the precarious economic balance of forest management, emphasizes the importance of presenting diversified, ecologically sustainable and economically attractive management alternatives. This implies a change in the traditional management focus, which is almost exclusively directed towards production and valuation of timber and underestimates the direct or indirect benefits that society gains from the fulfilment of other forest functions.

However, the above only makes sense if management is based on good planning, since increasing the interest in and demands on the forest will inevitably increase the likelihood of conflictive situations. For this reason, planning should allow a definition of the demands, plan for possible conflicts and offer adequate solutions, providing the other components of forest administration (forest policy, silviculture, forest economics, timber extraction and so on) with the basic information to define their objectives, take decisions and control their execution.

ANALYSIS OF FOREST FUNCTIONS AND THE RESPECTIVE IMPLICATIONS FOR FOREST MANAGEMENT

One reason that traditional forest management has failed as a valid tool for forest conservation is that it focuses almost exclusively on production and valuation of timber and not on the fulfilment of all the other forest functions, underestimating their direct and indirect benefits for society. For this reason, the following analysis attempts to emphasize the importance of forest management based on a holistic view, directed towards the fulfilment and valuation of all the different functions of the forest. To this end, an evaluation is made of the importance a forest may have for the fulfilment of each separate function, specifying examples of the implications that arise for forest management. It must be mentioned that any one forest does not necessarily have to fulfil all the functions analysed: there may be forests that only fulfil one of the functions mentioned, while others fulfil all of them. Nevertheless, this decision cannot be taken until a detailed analysis of the situation has been made, considering the elements that characterize each function separately. It should also be noted that, in order for forest management to guarantee fulfilment of the functions, the guidelines should be successively followed in the detailed planning of each forest management component (silviculture, utilization, economic aspects, infrastructure and so on). Some examples of the implications for forest management that result from different ecological functions of forests are presented below, with special reference to highland oak forests in Costa Rica.

Biodiversity Conservation

Most forest ecosystems house endemic species of flora and fauna. For this reason, it is indispensable to include conservation measures for these species in forest management, paying particular attention to any that are under some form of protection or are of special interest because of their ecological function. Selection may be made on the basis of forest typology (composition and structure) using legislative regulations for conservation of flora and fauna or taking into consideration the recommendations of conservation groups. In addition to listing the species, their characteristics and ecology should be studied and be taken into consideration when designing silvicultural interventions (Table 1.1).

In silvicultural study plots in the Villa Mills Demonstrative Experimental Area, these recommendations were applied and very promising results were obtained. In this zone, *Quercus costaricensis* can be considered the most important species, reaching values for absolute abundance of approximately 30 per cent and values for absolute dominance between 60 and 65 per cent (for both plots, before or after intervention). By contrast, due to the typology of the forest, *Quercus copeyensis* is never or hardly ever present. Of the 'Other Species' group, *Ilex* sp., *Ocotea* sp. and *Cythea gracilis* are the most abundant. As can be seen in Table 1.2, both the 20 per cent intervention (leaving 28.15m^2/ha) and the 30 per cent basal area (leaving 24.37m^2/ha) kept the same species that were present before interventions, altering the number of trees per species to an acceptable extent.

Table 1.1 Recommendations for forest management designed to fulfil the function of flora and fauna protection

Recommendations	Forest management component
Favour natural regeneration of *Magnolia* sp., *Podocarpus* sp., *Cythea* sp., locally known as 'chumico'	Silviculture
Favour species such as *Ocotea* sp., important as Quetzal (Pharomachus mocinno) food sources	Silviculture
Leave biological corridors as refuges for fauna	Soil use planning
Leave non-commercial timber on the ground as corridors for small rodents	Forest extraction
Leave dead trees standing for Quetzal nesting sites	Forest extraction
Apply careful extraction techniques to reduce impact on remaining stand[1]	Forest extraction
Promote natural regeneration of extracted species	Silviculture, forest extraction

For both treatments a new species appears in the post-impact inventories for trees with diameter at breast height (DBH) greater than or equal to 10cm carried out four years after silvicultural intervention. From the measurements of these trees it can be deduced that plants of this species were already present before intervention but were not included since their diameter did not reach the minimum (10cm) required for the inventory. For the same reason, it can be seen that in some cases (for example, *Dendropanax querceti*, *Weinmannia* sp., and so on) the number of trees for that species has increased four years after the intervention. It should be emphasized that a reduction in the number of trees per species is mainly due to silvicultural interventions and, to a much lesser extent, to natural tree death. Figures 1.1 and 1.2 show the impact of interventions on basal area (BA) distribution for each diameter class. In both interventions, basal area reductions were not concentrated only in the greatest diameter classes (as is the case for traditional forest utilization) since trees from smaller diameter classes were also cut, according to the needs for each case, in order to favour better quality trees or those of high ecological value.

On the basis of this analysis, it can be confirmed that the silvicultural interventions, although reducing the number of trees, left the forest structure fairly unchanged and did not change the species composition.

Table 1.2 Number of trees (N/ha) by species, before and 4 years after silvicultural intervention (DBH greater than 10 cm)

Species	Pre-impact 20% BA	Post-impact 20% BA	Pre-impact 30% BA	Post-impact 30% BA
Dendropanax querceti	14	16	7	22
Weinmannia sp.	19	10	14	16
Ilex sp.	62	46	41	38
Drimys granadensis	15	5	11	7
Cyathea gracilis	27	15	29	17
Podocarpus sp.	13	10	8	8
Quercus costaricensis	157	130	161	140
Ocotea sp.	59	33	70	55
Zanthoxilum sp.	13	8	15	8
Prunus cornifolia	29	13	24	19
Prumnopitys standl.	2	2	1	1
Vaccinium consang.	5	2	6	3
Saurauia costaric.	79	43	39	32
Schefflera pittieri	8	1	11	4
Styrax argenteus	17	15	11	11
Cleyera theaeoides	24	13	2	2
Quercus copeyensis	0	1	5	4

Soil Protection

Because of their structure, deep root systems, understorey and organic soil layer, natural forests are considered the best protection for soils, and in many cases the only feasible ones (Stadtmüller, 1988). Forests have a desirable effect on soil texture, structural stability, organic matter content, the type of clay minerals and soil permeability (Stadtmüller, 1994), so determining their susceptibility to erosion. Above all, in areas with steep slopes or high precipitation, the protective function on soils should receive great importance (Table 1.3).

In the experimental plots, the relatively high values for the number of trees with diameters over 10cm (Table 1.2), high level of natural regeneration (Table 1.4) and uniform distribution over the entire area show that good forest cover was maintained over the whole intervened area. It should also be mentioned that haulage operations were not carried out during the rainy season, trees within 20–30m of streams or water sources were not cut (if

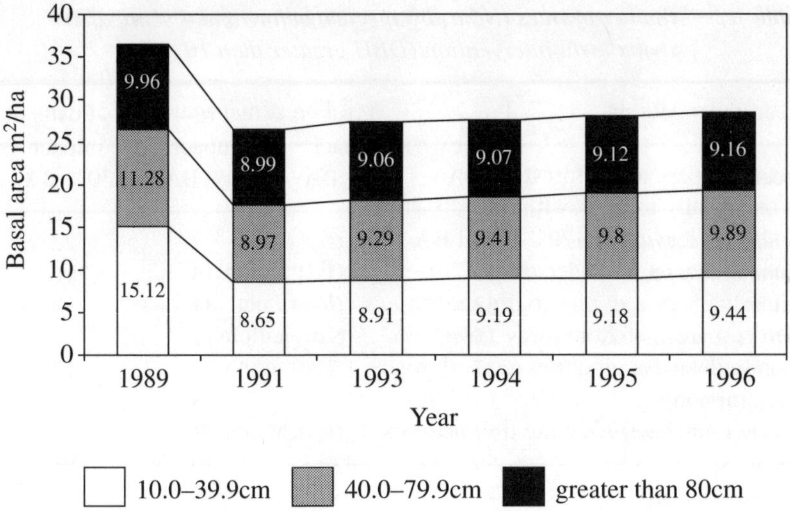

Figure 1.1 *Distribution of basal area by diameter class: intervention 20% basal area*

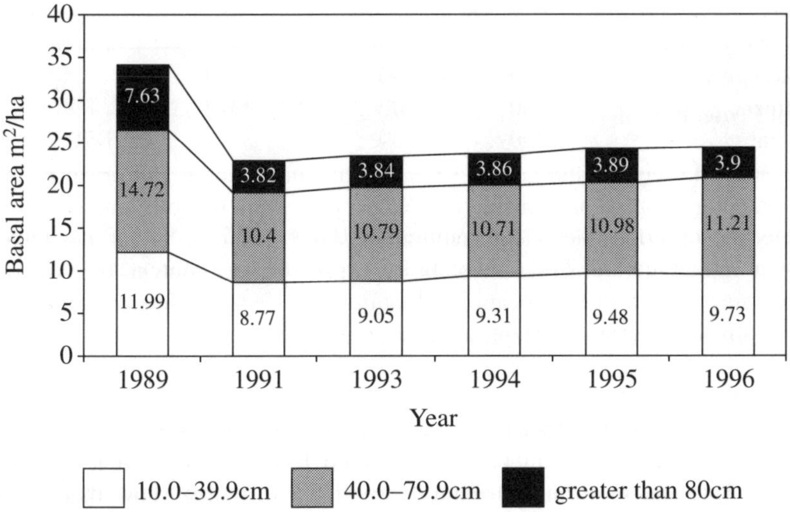

Figure 1.2 *Distribution of basal area by diameter class: intervention 30% basal area*

Table 1.3 Recommendations for forest management guaranteeing soil protection

Recommendations	Forest management component
Maintain permanent forest cover (especially in areas with over 30–40% slopes) leaving 60–70% initial BA	Silviculture
Reduce impact on understorey	Extraction
Limit haulage activities to dry season	Work plan, extraction
Reforest areas without forest cover	Silviculture
Plan in detail logging trail network for extraction	Extraction
Prevent machinery leaving trail network	Extraction

Table 1.4 Plant density by species and regeneration category (individuals between 0.30m and 9.9cm DBH), after silvicultural interventions

	Interv. 20% BA		Interv. 30% BA	
Species	Seedling (N/ha)	Sapling (N/ha)	Seedling (N/ha)	Sapling (N/ha)
Q. costaricensis	770	400	2 690	1 010
Q. copeyensis	0	30	0	0
Others	580	300	1 020	890
Total	1 350	730	3 710	1 900

selection criterion is only for commercial effects) and extraction machinery never left the trail network, reducing impact on the understorey and soil.

Conservation of Water Regime for Watersheds

Watersheds covered by forests are characterized by production of high-quality water, due to the forest's great capacity for high filtration rates (Stadtmüller, 1994). One study carried out in highland oak forests (Talamanca Cordillera, Costa Rica) showed that the water balance in a sustainably managed forest with intervention of 20–30 per cent basal area gave greater annual aquifer recharge (9300-10 500m^3/ha) than a forest without intervention (7600 m^3/ha) (Turcios, 1995). Although, to date, forest management receives no income to

Table 1.5 Recommendations aimed at fulfilling the function of water regime conservation in watersheds

Recommendations	Forest management component
Avoid haulage operations during wet season	Work plan
Avoid felling within 20–30m of streams or water sources	Silviculture
Avoid machinery leaving the trail network	Extraction
Avoid channelling trail network drainage directly into streams	Extraction

guarantee the fulfilment of this important forest function of water filtration, measures must be taken to ensure its fulfilment (Table 1.5).

Rain interception by tree crowns is one of the major influences on aquifer recharge in forest soils (Turcios, 1995). Consequently, reducing the number of trees in the forest reduces rain interception, leading to greater aquifer recharge. In this way, silvicultural interventions allow maintenance or increase in forest water production, as long as the basal area of the remaining trees does not limit the forest's role in protecting soil against landslides and erosion. To ensure minimal hydrological impact, Stadtmuller (1994) recommends not removing more than 30–40 per cent of the basal area. The same author also recommends that the intervention system should aim for a mosaic-type pattern, to ensure that possible temporary erosion foci are immediately buffered by surrounding non-intervened zones.

In the case of silvicultural interventions carried out on study plots in the Villa Mills experimental area, these and other measures recommended by Stadtmüller (1994) were applied to mitigate the hydrological impact. Greater details of monthly variations in the water balance, such as on the economic valuation of this forest management component, can be found in Turcios (1995). This demonstrates that forest management is not *a priori* incompatible with the conservation of the hydrological regime of a watershed, and may even improve its productivity (ibid.). It should be noted that the foregoing only refers to water production in quantitative terms and these results should be complemented by a qualitative analysis of the water produced.

Employment Income Generation

In order to evaluate the importance of forests with respect to employment income generation, their importance in satisfying basic needs such as fuelwood and charcoal, medicinal and ornamental plants, handicrafts and carpentry, poles and posts and so on must be analysed. The implications for forest management arising from these social functions greatly depend on the timber and non-timber products to be utilized, and a specific plan for their utilization is needed.

There are many forest management-related activities that allow involvement of the rural population, including women and children. These include production of fuelwood, fence posts and charcoal (using branches or twisted pieces from trees already utilized) and utilization of non-timber plant products such as fibres, edible, medicinal and ornamental plants, and so on, as shown in Table 1.6.

Table 1.6 Recommendations for forest management aimed at fulfilling the function of employment income generation

Recommendations	Forest management component
Train workers in the area in silviculture and extraction techniques	Work plan, training
Utilize branches of felled trees to satisfy demands for fuelwood and charcoal	Extraction, integrated utilization
Inventory biological diversity	Training, planning
Utilize the forest for ecotourism	Training, promotion
Design, manufacture and market handicrafts, souvenirs	Training, marketing
Involve women and youngsters in collection and marketing of other forest products	Extraction, financial analysis, social component

In the Villa Mills Demonstrative Experimental Area, both the felling and logging operations should only be carried out by well-trained local forest workers. (As shown in Table 1.9, interventions at 20 and 30 per cent basal areas allowed small farmers to produce 156 and $122m^3$ fuelwood per hectare and 337 and 220 posts per hectare, respectively.) These options ensured that the forest fulfilled an important function in generating employment, training and contracting local forest labourers, involving small farmers in the production and marketing of fuelwood and posts, and forming a forest producers association specializing in production (for charcoal), packing and marketing (formed by a group of women) that currently holds a working capital of approximately five million colones.

Recreation and Tourism

Over the last few years, ecotourism has become one of the main sources of income for Costa Rica. Since natural forests are an attraction for national and foreign visitors, this social function should not be ignored when planning forest management. (See Table 1.7.)

Table 1.7 Recommendations aimed at fulfilling the function of recreation and ecotourism

Recommendations	Forest management component
Construct interpretive paths (ecology and forest management aspects)	Planning, infrastructure
Construct bridges over streams for hikers	Infrastructure
Offer tours to hotels/tour operators	Work plan
Train local people as tour guides	Work plan, training, social, component

Production of Timber and Other Non-timber Plant Products

An integrated evaluation of the benefits that can be derived from diversified forest management cannot ignore the function of ecosystem production. It should be taken into account that forest production refers not only to the extraction of logs, but also to the utilization of residues for and the management of non-wood plant products present in the forest. A preliminary study, based on a compilation of local knowledge in an area of Costa Rican highland oak forest (Lafranchi, 1996) gave rise to a list of over 70 plant species used for different purposes. With this in mind, the planning process should include a compilation of information regarding the market for such products, aspects concerning operations for their exploitation and capacity of the forest farm including (financial potential of the farm, available equipment and labour. (See Table 1.8).

Guaranteeing timber production and sustainability

One of the great advantages of highland oak forests for silvicultural treatment is the acceptance in the market of all species that make up the upper canopy (although at limited prices). Furthermore, both *Quercus copeyensis* and *Q. costaricensis* are much quoted for fuelwood and charcoal production both in national and international markets. Because of these comparative advantages,

Table 1.8 Recommendations for forest management aimed at fulfilling the function of production

Recommendations	Forest management component
Draw up inventory of timber products available and respective marketing[2]	Planning
Enable local workers to learn better methods of extraction[3]	Training
Draw up plan for forest utilization: construct trail network	Extraction
Define measures for monitoring and control of operations and subsequent forest development	Administration
Make use of *medicinal plants* (*Weinmannia* spp., *Q. costaricensis*, etc.); *edible plants* (*Begonia* spp., fungi) and *ornamentals* (*Cythea* sp., *Sphagnum*)	Utilization, social aspects
Indicate estimate of initial investment and expected medium-term income	Financial analysis

Table 1.9 Timber products extracted through silvicultural interventions

Intervention intensity	Timber for sawmills (m^3/ha)	Fuelwood vol. (m^3 stere/ha)	Number posts (n/ha)
20% BA	36.1	156.1	337
30% BA	54.1	121.7	220

all trees felled for silvicultural interventions could be completely utilized, with trunks used for logs and branches used for fuelwood, fence posts, charcoal and supporting squash plants (Table 1.9).

The fact that more fuelwood was produced in the 20 per cent basal area (BA) intensity plot than in the 30 per cent plot is probably due to a greater percentage of felled trunks that failed to comply with the requirements of the national market for quality or size. These were used for fuelwood instead.

There follows an analysis of some important indicators to ensure sustainability of production. First, we consider the impact of silvicultural intervention on natural regeneration. Table 1.4 presented the values obtained for density of natural regeneration. For *Quercus costaricensis*, the density observed after intervention was considered adequate for future development assuring sustainability of forest management. However, in the case of *Quercus copeyensis* natural regeneration, the results were not so favourable. Nevertheless, this was not due to the silvicultural interventions themselves but rather to the type of forest used (mixed oak forest) in which *Q. copeyensis* is naturally relatively scarce. The species of ecological importance for this type of forest (*Cyathea gracilis* and *Podocarpus* sp.) were found at levels of 390 and 20 seedlings/ha, respectively, in forests with 20 per cent BA intervention and 240 and 110 seedlings/ha in forests with 30 per cent BA intervention. These values can be considered very promising for long-term conservation of these species.

With respect to post-intervention *mortality* (between January 1994 and January 1995), the plot with 20 per cent BA interventions showed lower survival percentages than the plot with 30 per cent BA intervention, both for *Q. costaricensis* and the 'Other Species' group. As regards *diameter increase*, Table 1.10 shows mean, maximum and minimum values for periodic diameter increases, along with annual increase for the total period, obtained during the first three years of measurements (1993–6). Negative minimum values are mainly due to mechanical damage (in the case of height) and climate change (in the case of diameter). For statistical analysis, mean values were used without the inclusion of the possible negative values mentioned. As can be seen, the forest with 20 per cent BA intervention showed greatest mean diameter increases during the year (1993–4) both for *Q. copeyensis* and the 'Other Species' group, with 0.7 and 0.2 cm/year, respectively. For the following period (1994–5) growth showed a slight reduction (0.1 cm for both groups). *Quercus costaricensis* showed no differences in diameter growth for the two periods (mean 0.2 cm/ year) and did not show a reduction until the period 1995–6.

Application of the Mann–Whitney test to compare diameter increase over the whole period for a single species under two treatments showed significant differences both for *Q. costaricensis* and the 'Other Species' group. Greatest growth was found in the 30 per cent BA intervention plot in both cases.

Results obtained for *increase in height* (Table 1.11) showed that growth increased from period to period. As in the case for diameter increase, significant differences were obtained for total height increase for the two treatments, greatest growth being shown in the plot with 30 per cent BA intervention.

From this analysis of the reaction of natural regeneration to the silvicultural interventions, it can be concluded that, given the density and high survival rate, natural regeneration after the interventions is sufficient to ensure management sustainability. In addition, it can be confirmed that greater diameter and height increases were found in the 30 per cent BA intervention for all species.

Table 1.10 Periodic diameter increase (cm) per species group and treatment (individuals between 0.3m height and 9.9cm DBH)

		Diameter Increase							
		20% BA intervention				30% BA intervention			
Species	Period	N	Mean	Max	Min	N	Mean	Max	Min
Quercus costaricensis	93–4	86	0.2	0.9	–0.2	148	0.3	1.0	–0.1
	94–5		0.2	1.0	–0.2		0.2	1.0	–0.2
	95–6		0.1	0.8	–0.1		0.2	0.8	–0.2
Quercus copeyensis	93–4	3	0.7	1.0	0.3	0	—	—	—
	94–5		0.3	0.6	0.0		—	—	—
	95–6		0.1	0.8	0.0		—	—	—
Other species	93–4	52	0.2	0.7	–0.1	98	0.2	1.0	–0.2
	94–5		0.1	0.5	–0.2		0.2	0.7	–0.2
	95–6		0.1	0.5	–0.2		0.1	1.0	–0.2

Table 1.11 Periodic height increase (m/year) by species and treatment (individuals between 0.3m height and 9.9cm DBH)

		Height Increase							
		20% BA intervention				30% BA intervention			
Species	Period	N	Max	Mean	Min	N	Max	Mean	Min
Quercus costaricensis	93–4	86	1.2	0.17	–0.48	148	1.22	0.25	–0.80
	94–5		1.4	0.24	–0.43		1.98	0.29	–1.00
	95–6		1.4	0.25	–0.70		1.85	0.25	–1.20
Quercus copeyensis	93–4	3	0.4	0.17	0.14	0	—	—	—
	94–5		0.3	0.20	0.15				
	95–6		0.3	0.30	0.15				
Other species	93–4	52	0.9	0.09	–0.93	98	1.18	0.18	–0.14
	94–5		1.3	0.11	–0.45		1.15	0.17	–1.20
	95–6		1.2	0.09	–0.88		2.40	0.20	–2.40

A second important indicator is the impact of silvicultural interventions on the remaining stand (DBH greater than 10 cm). *Tree mortality* (Table 1.12) was calculated on the basis of the formula in Swaine *et al.* (1987):

$$m(\%) = (\ln n_0 - \ln n_1) * 365 / t$$

where n_0 and n_1 are the number of trees registered in the first and second inventories, respectively (without including recruits) and t is number of days between one inventory and the next.

Table 1.12 Natural mortality rates for intervention intensity for the period before (1989–91) and after (1991–95) silvicultural interventions

	Natural mortality rate (%) per year	
Period	20% BA intervention	30% BA intervention
1989–1991	0.2	0.1
1991–1993	0.4	1.4
1993–1994	1.4	1.4
1994–1995	0.1	0.3

The results obtained indicate an increase in mortality after silvicultural interventions, but these values stayed relatively low compared to values indicated by Alder (1995) of 1.5 per cent for non-intervened forest and 2.5 per cent for forest which has been altered.

Recruitment is understood to cover those trees which between inventories enter the lower-diameter class (DBH = 10–39.9 cm). Values obtained are presented in Table 13 as n/ha/year (taking all species into account).

Table 1.13 Recruitment per hectare and year, before (1989–91) and after (1991–95) silvicultural interventions

	Recruitment (n/ha) per year	
Period	20% BA intervention	30% BA intervention
1989–1991	8	11
1991–1993	8	7
1993–1994	11	17
1994–1995	11	6

These values, although relatively low compared to those obtained in Malaysia by Swaine *et al.* (1987) of 11.8 n/ha/year (1947–63) and 20.98 n/ha/year (1971–81), are considerably greater than the mortality rates found.

Table 1.14 *Average and variance of diameter increase for species group, diameter class and intervention intensity, for the periods 1989–91 and 1991–5.*

(a) *Quercus costaricensis*

			Pre-intervention increase (cm/year)		Post-intervention increase (cm/year)					
			1989–1991		1991–3		1993–4		1994–5	
Intervention intensity	Diametric class	(N/ha)	Average	Var	Average	Var	Average	Var	Average	Var
20% BA	10.0–39.9cm	79	0.44	0.09	0.54*	0.13	0.52	0.15	0.43	0.11
	40.0–79.9cm	23	0.62	0.17	0.52	0.14	0.57	0.26	0.49	0.13
	> 80cm	6	0.38	0.11	0.43	0.24	0.60	0.12	0.29	0.07
30% BA	10.0–39.9cm	76	0.47	0.08	0.58*	0.12	0.65*	0.18	0.58*	0.14
	40.0–79.9cm	28	0.48	0.10	0.43	0.15	0.40	0.16	0.36	0.09
	> 80cm	3	0.23	0.42	0.11	0.01	0.31	0.42	0.49	0.30

Note: *Increases significantly higher than pre-impact increases (t-test, C 95 per cent).

Table 1.14 shows a summary of the impact of silvicultural interventions on *diameter increase* for *Q. costaricensis* trees, comparing the values obtained before intervention (pre-impact) and after (post-impact).

As can be seen, the greatest impact of the interventions was mainly observed in *Q. costaricensis* trees in the diameter class 10–39.9cm. In the case of 30 per cent BA intervention intensity, the favourable impact on diameter increase remained until the last measurements were carried out (1994–5). However, in the case of 20 per cent BA intervention, statistical evidence of greater diameter increase was only found for the period following the interventions. On the basis of the results obtained in this study, an analysis was made of a possible correlation between annual diameter increase and the following factors:

Quercus costaricensis:
Crown illumination

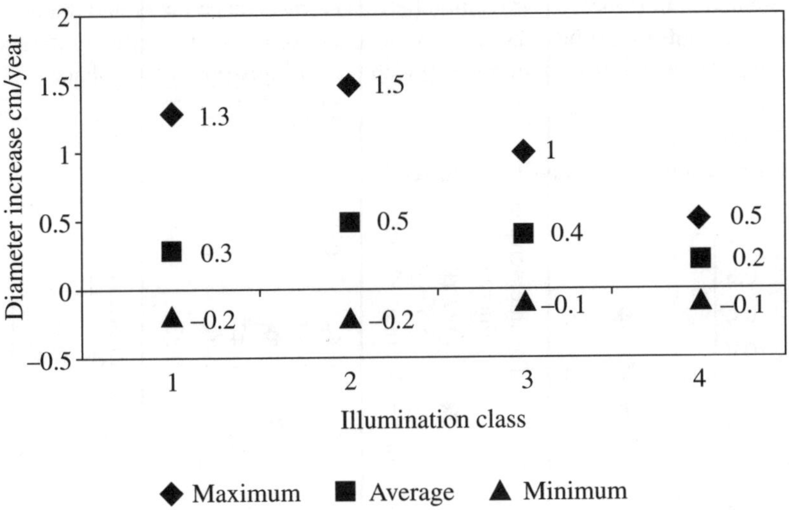

Notes: Illumination Classes Hutchinson (1993), adapted from Dawkins, Silvicultural Research Plan, Uganda, 1959–63:

1 Full vertical illumination, along with lateral.
2 Full vertical illumination.
3 Partial vertical illumination.
4 Oblique illumination only.
5 No direct illumination.

Figure 1.3 Annual diameter increase (cm/yr) as a function of illumination class and crown shape

- Crown illumination (Dawkins' classification, adapted by Hutchinson, 1993.
- Crown shape (adaptation of Dawkins' classification, cited by Blaser, 1987.

The results of this analysis for the species *Q. costaricensis* and *Q. copeyensis* are shown in Figures 1.3 and 1.4 (data from other plots in the same study were also used for the purpose of this analysis).

Mean values for diameter increase in *Q. costaricensis* indicate that this species exhibits a degree of flexibility with regard to crown illumination. Greatest increases were obtained in trees with crowns that were completely free above. Nevertheless, very satisfactory increase rates were obtained both for trees with completely free and for those with partially free crowns. However, trees with crowns that were partially or completely covered showed a marked (statistically significant) lowering in growth increase. The factor 'crown shape' seemed to have an even greater influence on diameter increase. The data presented with respect to correlation between crown shape factor and diameter increase confirm that best results were obtained for trees with perfectly formed crowns. Diameter growth diminished with increasingly poor crown shape.

(b) *Quercus copeyensis*:
Crown illumination (classification: see Figure 1.4)

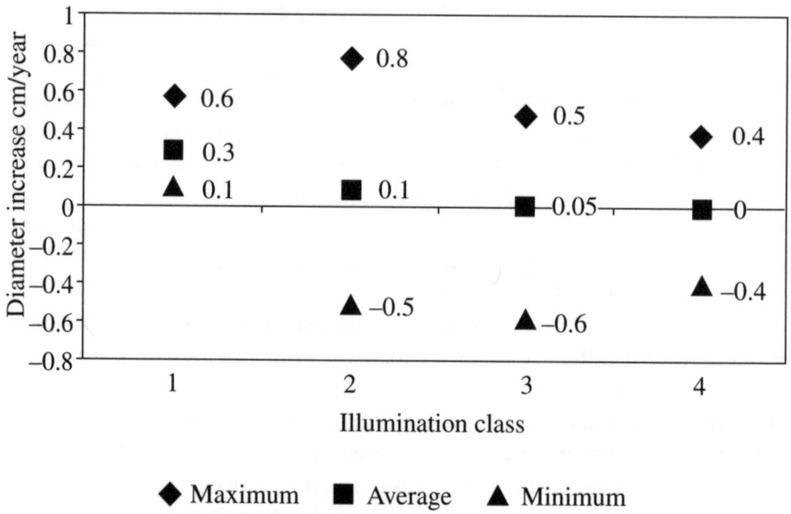

Figure 1.4 Annual diameter increase (cm/yr) as a function of illumination class and crown shape

Even with a completely free crown, *Quercus copeyensis* showed considerably lower values than *Q. costaricensis* for mean diameter increase. This species appears to have less flexibility with respect to crown illumination. Furthermore, the crown shape factor has less correlation with diameter increase than was the case for *Q. costaricensis* (Table 1.15).

Table 1.15 Spearman correlation factor between illumination class/crown shape and diameter increase

		Quercus costaricensis		*Quercus copeyensis*	
Period	Statistical indicator	Crown illumin.	Crown shape	Crown illumin.	Crown shape
1991–3	Spearman coeff.	–0.3	–0.01	–0.04	–0.06
	N	578	578	489	489
	Level of signif	0.0000	0.7489	0.4165	0.2074
1993–4	Spearman coeff.	–0.3	–0.12	–0.08	–0.19
	N	578	578	489	489
	Level of signif	0.0000	0.0027	0.0716	0.0000
1994–5	Spearman coeff.	–0.4	–0.1	–0.11	–0.20
	N	578	578	489	489
	Level of signif.	0.0000	0.0073	0.0149	0.0000

A comparison of the light conditions needed by these species and conditions found in the forest after a 30 per cent BA reduction intervention (Figure 1.5) shows that illumination conditions for the trees are favourable for their growth.

If, on the other hand, illumination conditions found in forests without silvicultural intervention are analysed (Figure 1.6), it can be seen that most trees that are found under low illumination do not achieve optimal growth. For this reason, silvicultural intervention is recommended.

As mentioned earlier, crown illumination conditions and crown shape are factors that most influence diameter increase in trees. Silvicultural interventions have aimed to improve forest conditions with respect to these parameters. However, in primary forest, this objective can only be achieved over the long term, gradually improving conditions after each intervention. On this basis, and taking into account the fact that the first interventions carried out in the study plots mainly concentrated on the removal of badly formed trees (rather than favouring good-quality trees), it is not surprising that the forest's response is still only moderate.

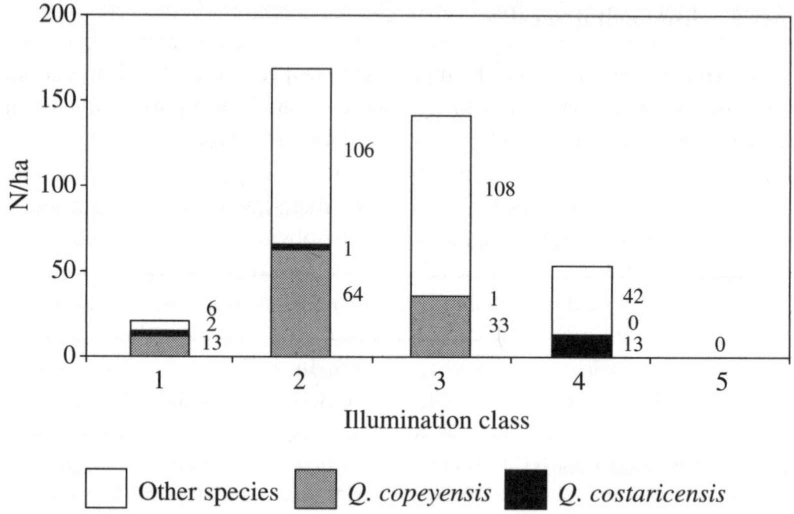

Figure 1.5 Illumination condition of trees (DBH greater than or equal to 10cm), 30% BA intervention

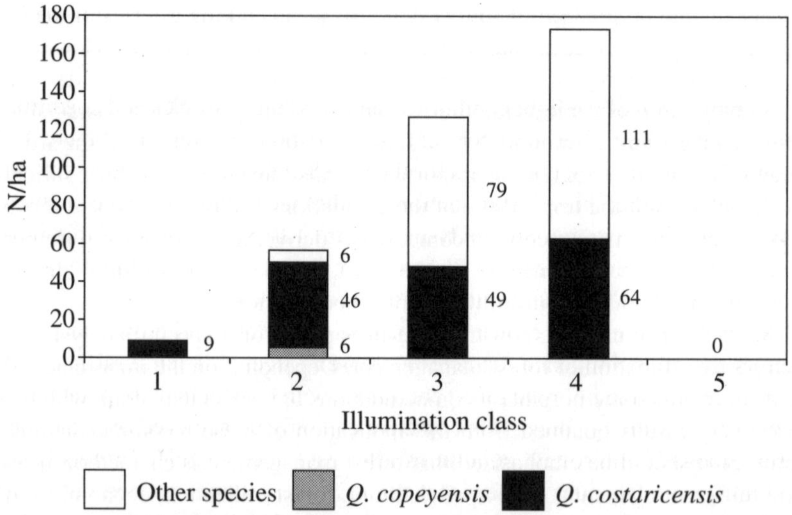

Figure 1.6 Illumination condition of trees (DBH greater than or equal to 10 cm), without intervention (50.52 m^2/ha)

Other Environmental Services

Another forest function recognized at world level is the ability of forests to act as carbon sinks. The suggestion that vegetation is an atmospheric carbon sink is supported intuitively by past accumulations of fossil fuels originating from photosynthesis and the continued tendency of vegetation to accumulate carbon (Lugo and Brown, 1992). Scientific debate about the role of tropical forest in the global carbon cycle transcends the obvious difficulties caused by poor databases (ibid.).

Despite the lack of information, it has been argued that forest ecosystems conserve 20 to 100 times more carbon per unit area than agricultural ecosystems (Andrasko, 1990, cited by Ortiz, 1996). In the case of oak forest, the high number of trees per hectare, equally high volumes of timber and depth of the soil organic layer lead one to expect that they fulfil an especially important function in the fixation and storage of carbon. Studies are currently being carried out in the Villa Mills Experimental Area to evaluate carbon fixation and storage in these ecosystems (Segura, 1997).

CONCLUSIONS AND RECOMMENDATIONS

An analysis of forest functions and their respective implications for forest management reveals that forest management does not merely mean ensuring the sustainable production of timber, but also means taking a series of measures to ensure the fulfilment of other intrinsic forest functions. However, despite the importance of these functions, it has not been possible, to date, to assign them an economic value that would contribute to making sustainable forest management economically attractive for forest conservation. In order to achieve this, it is extremely important to raise awareness at the political and social level of the importance of these functions and of the rights of those who manage forests to receive income that is derived not solely from timber production (which, furthermore, is the most destructive operation of forest management) but also from fulfilment of other functions.

This chapter aims to demonstrate the importance of a specific analysis for each possible function of forest management. Depending on the importance of these functions, some possible recommendations are presented, along with the preliminary results obtained from the application of these recommendations. However, it should be emphasized that forest management is characterized by many different elements, such as silviculture, forest utilization, economic and financial aspects, social aspects and so on, which need specific and careful planning. The recommendations set out in this analysis should, therefore, be reviewed separately for each management component during specific planning.

Depending on the ecological, social, economic and cultural characteristics, forest functions will vary considerably. This is another reason why a detailed analysis should be made for each specific case. Forest management is confined within paradigms which, to date, have limited the social and economic benefits that can be derived. There should be social benefits derived from the optimum fulfilment of all intrinsic forest functions and economic benefits, above all for forest owners offering the remuneration deserved for integrated forest management. The main drawback is perhaps rooted in the idea of seeing the value of the forest solely in terms of the volume of exploitable timber. Instead of concentrating simply on monetary costs and benefits, it is time to look at possible future *threats* and *opportunities*.

NOTES

1. Studies carried out in the experimental area of Villa Mills (Talamanca Cordillera, Costa Rica) showed that effects on rodents and birds are minimal if silvicultural intervention is based on felling selected trees, applying improved utilization techniques (extracting between 20 and 30 per cent of basal area).
2. On the basis of recommended intervention intensity (20–30 per cent basal area), results from research carried out in the Villa Mills experimental area (Talamanca Cordillera, Costa Rica) for highland oak forest indicate that the following approximate production can be expected: $60m^3$/ha logs for sawmills (mainly Quercus spp.), $170m^3$/ha steres fuelwood and 340 posts/ha used as trellis for growing 'chayote', a local squash. Considering estimated growth rates for highland oak forest, these interventions should be made every 15 years. If the time interval between interventions is increased or decreased, the intensity of intervention should be modified accordingly, although increased intensity is not recommended even in the case of longer intervals between interventions.
3. Training should cover subjects such as silviculture, planning, construction and maintenance of trail network, logging techniques, first aid and so on.

REFERENCES

Alder, D. (1995), 'Growth Modelling for Mixed Tropical Forests', Oxford Forestry Institute, Dep. of Plant Sciences, Tropical Forestry Paper **30**, University of Oxford, 231.

Blaser, J. (1987), 'Standörtliche und waldkundliche Analyse eines Eichen-Wolkenwaldes (*Quercus spp*) der Montanstufe in Costa Rica', Tesis de Doctorado, publicado en: *Göttinger Beiträge zur Land- und Forstwirtschaft in den Tropen und Subtropen*, **26**, Göttingen, BRD, 235.

Hutchinson, I.D. (1993), 'Puntos de partida y muestreo diagnóstico para la silvicultura de bosques naturales del trópico húmedo', *Colección silvicultura y manejo de bosques naturales*, no. 7, CATIE, Serie Técnica, Informe Técnico, **204**, 32.

Lafranchi, S. (1996), 'Los productos vegetables no maderables en los robledales de altura y sus alrededores', Cordillera de Talamanca, Turrialba, Costa Rica, CATIE, 107.

Lugo, A.E.; Brown, S. (1992), 'Aboveground biomass estimates for tropical moist forests of the Brazilian Amazon', Serie: Interciencia (Venezuela) **17** (1), 8–18.
Ortiz, R. (1996), 'Bosque secundario por abandono de pastizales en la región Huetar Norte de Costa Rica: análisis de algunos criterios económicos y ambientales', Tesis de Maestría en Economía Ecológica, CINPE, Costa Rica.
Segura, M. (1997), 'Almacenamiento y fijación de carbono en Q. costaricensis en los robledales de la Cordillera de Talamanca, Costa Rica', Tesis de Lic, Universidad Nacional, Costa Rica, 134.
Stadtmüller, T. (1988), 'Funciones climáticas e Hidrológicas de los bosques con énfasis en los trópicos', El Chasqui, **16**, Costa Rica, 3–10.
Stadtmüller, T. (1994), 'Impacto hidrológico del manejo forestal de bosques naturales tropicales, medidas para mitigarlo', *Colección Silvicultura y manejo de Bosques Naturales*, **10**, Informe Técnico **246**, CATIE, Turrialba, Costa Rica.
Swaine, M.E., D. Lieberman and F.E. Putz (1987), 'The dynamics of tree populations in tropical forest: a review', *J. Trop. Ecol.*, **3**, 359–66.
Turcios, W. (1995), 'Producción y valoración económica del componente hídrico y forestal de los robledales de altura bajo intervenciones silviculturales', Tesis de Maestría, CATIE, Turrialba, Costa Rica, 97.

2. Integrating socioeconomic and biophysical data with a GIS to plan sustainable resource management efforts in Central America

Steven Shultz

INTRODUCTION

Sustainable resource management efforts in Central America and in many other tropical regions of the world are often focused on the watershed level of analysis and commonly promote soil conservation, agroforestry and reforestation techniques among small hillside farmers (Lutz *et al.*, 1995; Current *et al.*, 1995). Regularly needed information for planning and implementing such activities includes the identification of critical land areas, the selection of appropriate resource management practices and, finally, the estimation of the costs and benefits of such practices.

It is argued here that, for such resource planning and management efforts to be effective and sustainable over time, they must take account of biophysical, socioeconomic and institutional processes which together influence resource use, land use degradation and the adoption of natural resource management technologies by farmers. The importance of including socioeconomic data in the identification of critical areas within watersheds and protected areas is obvious: critical land uses are a result not only of biophysical characteristics of land areas, but of the actions, interests and behaviour of the people who live and work on these lands. More specifically, it is necessary to estimate the likelihood that local farmers and resource users will participate in soil conservation programmes, reforestation projects and/or natural forest management programmes. Similarly, in cases where the government intends to purchase critical lands in order to establish protected areas, it is necessary to know the corresponding production or property values in order to estimate the costs of government land expropriation programmes. Such socioeconomic data are usually collected via farm-level surveys and through agricultural census programmes.

The *a priori* determination of specific details and plans, required for efficient implementation of resource management projects, is a particularly important issue in Central America and other tropical developing countries that have many degraded watersheds and limited financial resources for natural resource management activities. That is, with so many degraded watersheds, and with such an acute scarcity of financial resources, resource managers need to be able to prioritize and characterize areas within watersheds most in need of restoration efforts.

The objective of this chapter is to promote GIS (geographic information systems) based methodologies for planning resource management activities in tropical watersheds while taking account of the typical resource, data and time constraints faced by many natural resource professionals in Central America, and in many other tropical regions. After a summary of the use of GIS technology for natural resource planning and management activities, two applied case studies are used to illustrate the utility of such technologies. In the first case study, GIS technology is used to integrate land use and farmer-level socioeconomic data in order to plan a reforestation effort for the highly degraded agricultural Purire River watershed of central Costa Rica. In the second case study, similar GIS-based methodologies are used to identify priority areas for protection and/or soil conservation, agroforestry and natural resource management programmes in the pristine multiple- use Pacuare River Watershed on the Atlantic slope of Costa Rica.

THE USE OF A GIS TO INTEGRATE SOCIOECONOMIC AND BIOPHYSICAL DATA

A Brief Review of GIS Technology

A GIS is a computer-based system to capture, store, edit, manipulate and display geographically referenced information. Rather than being considered to be simply an elaborate computer graphics or map-making program, GIS should be recognized as a tool that can produce high-quality graphics and maps, but also enables users interactively to model, manipulate and analyse many quantitative and qualitative features of the real world.

Most GIS systems incorporate the following three basic functions: data entry, manipulation and display. Data entry can involve digitizing (or scanning) spatial information (usually maps) into a computer with the aid of software or simply entering descriptive data that directly relate to spatial data (points, lines or areas). Various data layers (also called 'covers' or 'themes') containing both spatial and descriptive data, can be integrated, analysed and presented in a wide variety of ways as a result of data entry activities.

With selected data overlays (or themes) there are numerous types of data manipulations that can be carried out with GIS software, including logical, mathematical, geometric and network analyses, and a variety of other spatial and temporal modelling procedures. Advantages associated with utilizing GIS technology include the following:

- large volumes of data are maintained in a compact form;
- a variety of types of manipulation, including map measurement, thematic map overlays, transformations, spatial analysis and data base analysis, can be conducted relatively easily and quickly;
- graphic and tabular information (that is, attribute data) can be merged and manipulated in a related manner;
- the testing of conceptual models which represent spatially related data can easily be performed, which in turn facilitates both scientific and public policy analyses.

The Use of a GIS to Identify Critical Land Use Areas

Both the establishment of formal protected watershed areas and the implementation of soil conservation and agroforestry programmes for watersheds require the identification of critical areas within a specific watershed or study area. This involves the collection and analysis of biophysical data such as slopes, soils, hydrology, vegetative cover and land uses. Most often, these biophysical data are collected and analysed in a spatial and temporal context, through the use of GIS and satellite imagery.

The identification of critical land areas requires the collection and analysis of a variety of biophysical and land use data. This has become considerably easier in Central America and in other developing countries in recent years with the use of GIS technologies and the increased availability of a wide range of spatially referenced biophysical types of GIS data (Jimenez and Faustino, 1996).

The most commonly used GIS technique for identifying critical areas is the 'spatial overlay'. Specifically, this involves defining a spatial relationship between two or more data themes that share a common cartographic referencing system. Once such spatial relationships are established a variety of logical, mathematical (Boolean) or geometrical analyses can be jointly performed on the data in order to quantify the complex interrelationships between the data themes (Maguire, 1991). Such a spatial overlay is shown in Figure 2.1.

For example, it is possible to identify critical areas in a watershed by overlaying separate GIS data coverages of soils or geology, elevation, rainfall, land uses, vegetation and so on. The specific geographical areas containing various combinations of the above resource characteristics that make them

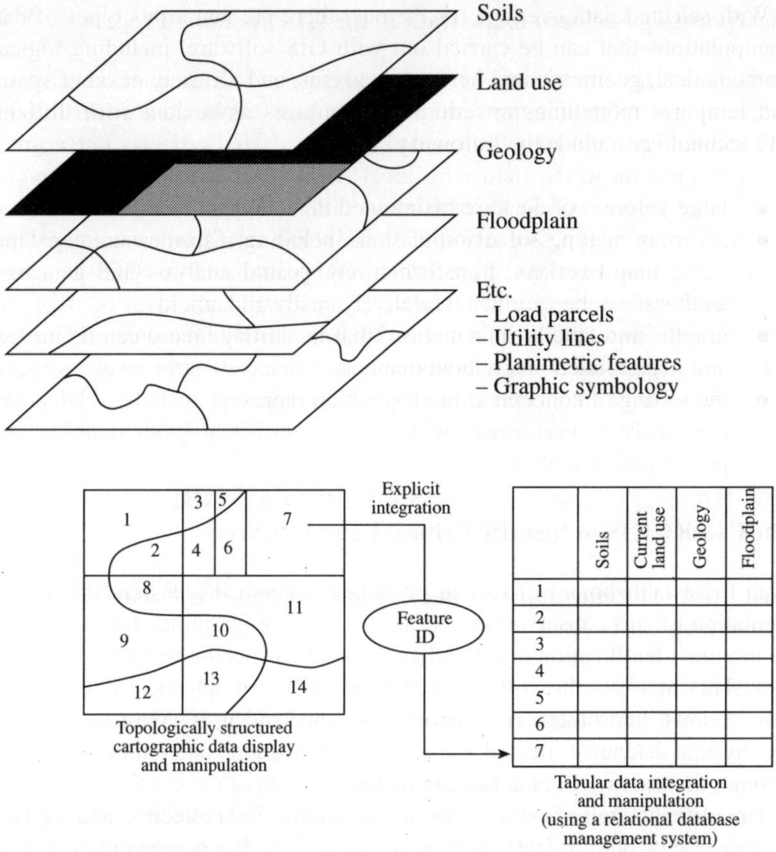

Figure 2.1 Example of spatial overlay functions of geographic information systems

particularly prone to high levels of land degradation and soil erosion can then be calculated, using a variety of selection criteria (Adinarayana *et al.*, 1994; Theocharoppolos *et al.*, 1995). In most cases, such critical land use areas are deforested lands being used for hillside agriculture or intensive cattle grazing activities. An equivalent spatial overlay approach to identifying critical land use areas is to overlay actual land use maps with land use capacity maps (which are themselves made based on soil, slope and land use suitability indexes), in order to identify areas of inappropriate or conflicting land uses.

Incorporating Socioeconomic Data in the GIS Planning Process

The key strategy or method to integrate socioeconomic data and biophysical data is by geo-referencing of farm locations. This can be done either through government cadastral based (property) maps, cadastral property line correcting and updating in the field with local farmers, or with the use of global positioning system (GPS) technology. A GPS is a small hand-held receiver which, by triangulating signals with space-borne satellites, accurately determines the spatial locations of specific ground point locations. GPS-generated data can therefore identify the real-world (spatial) locations of roads, farms and property lines, and these data can be immediately incorporated into GIS databases. Again, once farm boundaries are mapped, farm-level socioeconomic data can be integrated with biophysical data.

GIS-BASED CASE STUDIES OF SUSTAINABLE RESOURCE MANAGEMENT

Case Study 1: Reforestation Planning in the Degraded Purire River Watershed, Costa Rica

Introduction

The Purire River watershed is a sub-watershed of the Reventazon River on the Atlantic slope of central Costa Rica, located in the Guarco Valley, 5 kilometres southwest of Cartago, which is Costa Rica's second-largest city. It is a relatively small, humid, tropical watershed, 7500 hectares (ha) in size. The watershed is characterized by volcanic soils with low infiltration, a range in elevation between 1380 and 2100 metres above sea level, average annual rainfall of 1760mm (concentrated mostly in the rainy season from May to November) and an average annual temperature of 19°C. Adjacent tributaries of the Purire River in the watershed are the Cabrera and Coris rivers.

The upper part of the watershed is steeply sloped and classified as suitable only for forestry, yet it is mostly deforested and used primarily for cattle grazing. The lower part of the watershed is relatively flat, and suitable for various agricultural uses. It contains several rural communities, some cattle grazing and some-small scale vegetable production, but is mostly noted for its very intensive agricultural export industry based on growing ferns, flowers and other ornamental plants within screen houses. In 1989, approximately 280 hectares were dedicated to such agricultural export industries in this watershed, which employed more than 3000 people.

Throughout the 1980s the lower watershed was subject to a series of recurrent and severe flooding. Although no detailed hydrological studies were made, these floods have been attributed to the rapid deforestation of the watershed, influenced by cattle grazing subsidies provided by the government of Costa Rica in the 1970s. In December 1988, the watershed was declared to be in a state of national emergency from flooding, with costs of damage to homes and property, roads and bridges, and to agricultural export operations exceeding $650 000. In the aftermath of this disastrous event, many of the local community leaders and members of the association of agricultural export producers in the Purire watershed expressed interest in both short-term downstream structural flood control measures and medium to long-term reforestation efforts in the upper watershed. However, no agreements were ever reached regarding the extent and details of the reforestation efforts and, specifically, who should finance them. In addition, farm owners in the upper watershed showed no direct interest in implementing reforestation efforts themselves without external financial and technical assistance.

Immediately following the implementation of the structural flood control measures and the subsequent reduction of the threat of flooding, interest in upstream reforestation efforts lessened. However, in the last few years, after observing sharply increased needs (and costs) for maintenance activities associated with the structural flood control measures (due to increased levels of soil erosion and sedimentation), a renewed interest in reforesting the upper watershed again developed.

There still existed a great deal of uncertainty about how to implement a reforestation plan for the upper Purire watershed. In particular, the specific priority areas needing to be reforested were not known. It was not determined whether upstream farmers would want to participate in reforestation efforts, and the costs of the reforestation programme were unknown.

To resolve these uncertainties regarding the reforestation plan, the following activities were undertaken in 1995 by a team of graduate students and technicians at CATIE (the Tropical Agricultural Research and Higher Education Center), located in Costa Rica. First, critical land areas in need of reforestation were identified using GIS technology. Second, the preferences of local landowners were elicited through surveys in order to select appropriate forestry species. Finally, a potential reforestation programme was formulated and its economic feasibility evaluated (Medina *et al.*, 1997).

The identification of critical land use areas

The identification of critical land areas in need of reforestation involved using a GIS to overlay elevation, actual land use and land use suitability coverages. Elevation and soils coverages were obtained and digitized from 1:50 000 and 1:10 000 scale topographic maps and used to validate the quality of the land use suitability map, and to obtain specific information regarding soils and elevation in the watershed.

It was determined that approximately 30 per cent of the lands in the watershed can be considered flat, with slopes less than 12 per cent, while 48 per cent are moderately sloped (between 12 per cent and 50 per cent), and 12 per cent of the land in the watershed is very steeply sloped (slopes greater than 50 per cent).

Actual land use areas were digitized from 1:10 000 scale topographic maps of the region that included general land uses. These maps were constructed in 1991, based on 1989 air photos of the watershed, through a joint Japanese Development (JICA) and National Geographical Institute (IGN) project. These maps were then checked for quality control through field checks and comparisons with a more recent, but less detailed, satellite image of the area.

Actual land uses in the watershed are shown in Figure 2.2. Only about 870 ha, or 11 per cent, of the watershed remain in forest cover (combined primary and secondary forests). Secondary growth or charral (natural regeneration on abandoned pasture lands) and actual pasture lands are the most common land uses, with approximately 2500 ha each, which collectively represent 66 per cent of the total lands in the watershed.

A land use capacity map of the study area was obtained from the Neotropico Foundation. Fourteen land use categories were mapped, ranging from agricultural uses to protection based on a combination of factors including soils, elevation and distance to rivers and streams (CCT, 1994). This land use capacity mapping classification is quite standardized and has been made for much of central Costa Rica. For the purpose of the present study, this land use capability map was digitized into a GIS format and checked for accuracy through comparisons with other soil and elevation maps (Figure 2.2).

For purposes of a simplified display, Figure 2.3 portrays land use suitability in the watershed classified by just four general classes. Almost 25 per cent of the watershed (1890 ha) is deemed suitable for agricultural activities. This corresponds to the flat downstream parts of the watershed where most urban settlements are located, and where almost all of the intensive agricultural export crop activities take place.

About 30 per cent of the watershed (2220 ha) is considered suitable for forestry or permanent crops, while 24 per cent (1765 ha) is suitable only for forestry, and the remaining 21 per cent of the watershed (1635 ha) is recommended to be protected, with no agricultural or forestry extraction activities. The protected areas include approximately 616 ha of very steeply sloped land, and 1019 ha of buffer zones of between 20 and 100 metres on either side of the various streams and rivers in the watershed.

Critical land areas were identified by performing a GIS-based spatial overlay of coverages representing actual land use and land use suitability. Areas were determined to be in 'overuse' when actual land uses were more inten-

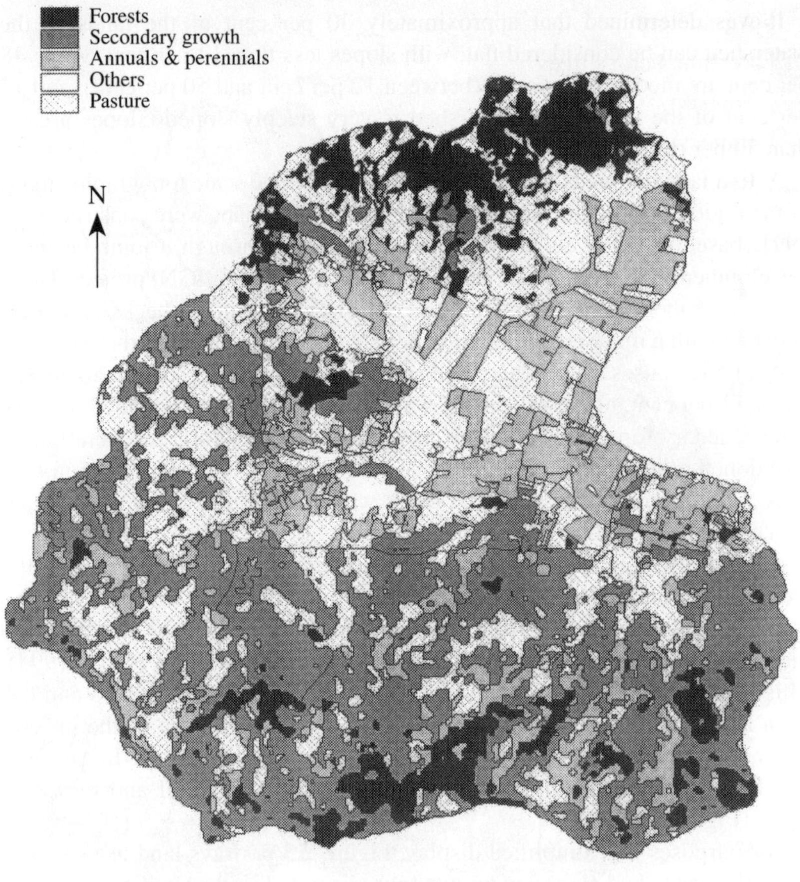

Figure 2.2 Land uses in the Purire River watershed

sive than land use capacities, in 'underuse' when actual land uses were less intensive than land use capacities, and in 'correct use' when actual land uses were generally equal to land use capacities. Urban land use, lakes, roads, mines and other miscellaneous land uses were not classified in terms of land use conflicts. Areas of land use conflict (the 'overuse' case) in the watershed are shown in Figure 2.4.

Approximately 2365 ha, or 32 per cent of the watershed, can be classified as overuse or critical land use areas. Underuse areas are only 435 ha, or 6 per cent of the watershed, while 60 per cent of the watershed (4472 ha) is being correctly used.

Finally, the labour of this GIS-based exercise was estimated to be 1.5 complete months (45 days). It can also be assumed that access to GIS-based

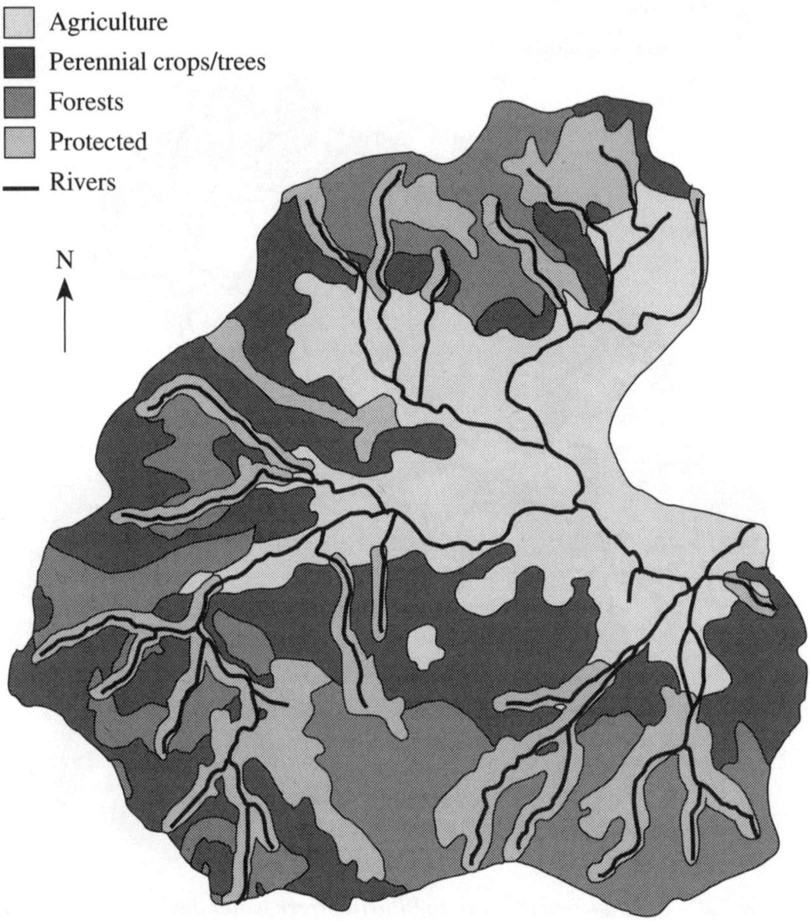

Source: Centro Cientifico Tropical (1994).

Figure 2.3 Land use suitability in the Purire River watershed

computational equipment with GIS software is also needed. Other required supplies include air photographs, and possibly satellite images. However, as the costs of GIS hardware, software, data and manpower are continuing to fail, these cost requirements are not considered prohibitive, even in a developing country setting.

As expected, most of the critical land use areas are located in the upper part of the watershed. Around 63 per cent of critical land use areas in the watershed (1490 ha) are pasturelands, located mostly in areas designated for

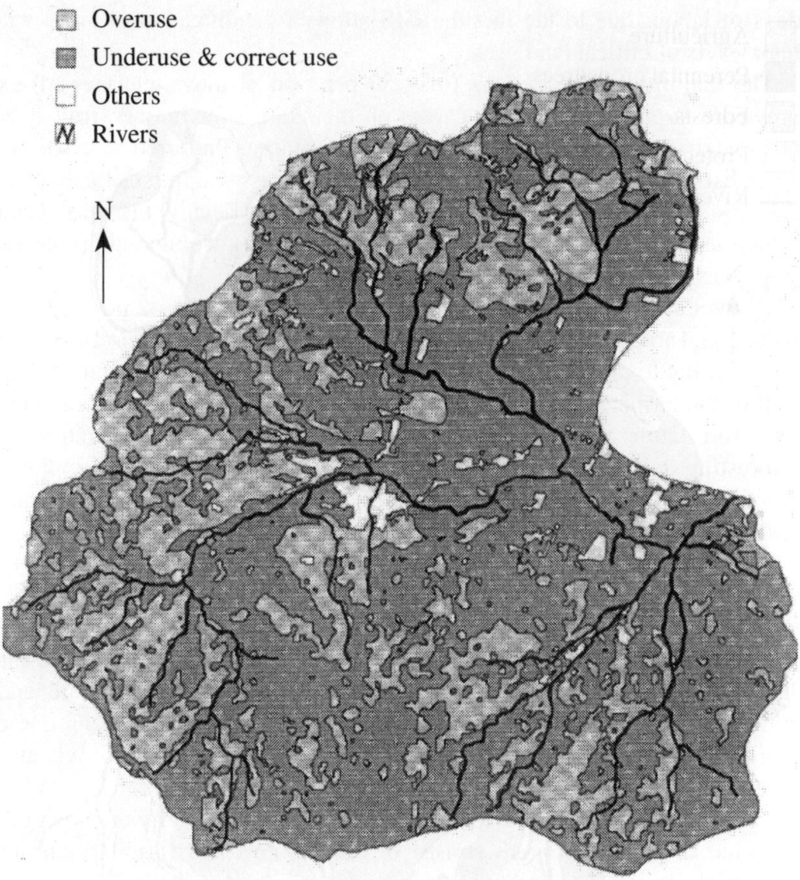

Figure 2.4 Land use conflicts in the Purire River watershed

forestry or protection, while 19 per cent (449 ha) of the critical lands are perennial or annual crops and, finally, 18 per cent of the critical land areas in the watershed (426 ha) are lands buffering rivers and streams by 20 metres in low-lying areas and by 100 metres in steep areas.

Incorporating farmers' preferences in the formulation of a reforestation plan

A series of personal surveys and interviews was conducted with a random sample of 16 farmers living within the identified critical land use areas of the upper watershed, in order to identify their preferences with regard to forestry species and planting systems. These surveys only took approximately three days of labour due to the fact the GIS allowed identification of areas with high levels of critical land uses.

The majority of the farmers (over 80 per cent of those interviewed) expressed a strong desire to plant trees on their farms for lumber, fruit trees, posts and firewood (listed in order of preference). Preferred tree species, again listed in their order of preference, were cypress (Cupressus Lusitanica), casuarina (Casuarina Equisetifolia), eucalyptus (Eucalyptus sp.), cedar (Quercus Costarricense), jual (Alnus acuminata), roble oak (Quercus Copeyensis) and Encino oak (Quercus Costarricense).

Of the farmers interested in participating in reforestation, 42 per cent were willing and able to reforest without any financial assistance, while 16 per cent were willing to provide labour and transport costs, and the remaining 42 per cent of the farmers stated they would only be willing to provide labour costs for reforestation activities. Farmers almost universally expressed interest in reforesting degraded pasture lands with block plantations and incorporating agroforestry-type systems (live fences, shade trees and so on) on their lands that were under some form of annual or perennial crop production.

Almost 80 per cent of the farmers within the critical land use areas of the watershed were willing and at least partially able to finance the reforestation of their lands. From this it was estimated that 1892 hectares of farmland in the watershed were available for reforestation. It was also determined that 1192 ha of these lands available for reforestation were degraded pasture lands and therefore deemed most appropriate for plantation forestry using one or more exotic or native tree species preferred by local farmers. Approximately 30 per cent (358 ha) of these lands designated for plantations consist of Class I soils, defined as favourable to tree growth, while the majority or 70 per cent (834 ha) of plantation areas consist of less fertile Class II soils with only moderate tree-growing potential.

Correspondingly, the 359 ha of critical land use areas where annual or perennial agricultural production is taking place were determined to be best suited for agroforestry systems such as live fences, windbreaks and taungya systems. Again the rationale for this is that farmers have made significant investments in these agricultural systems and for the most part stated that they are not interested in abandoning them for plantation forestry systems.

Finally, the 426 ha of critical lands buffering the streams and rivers of the watershed are assumed to be most effectively reforested through natural regeneration which, if left undisturbed, should provide a very effective protective cover to the most sensitive soils in the upper watershed. This would involve little or no cost for the farmers and would help justify the protected status of these lands. This proposed reforestation plan is illustrated by Figure 2.5.

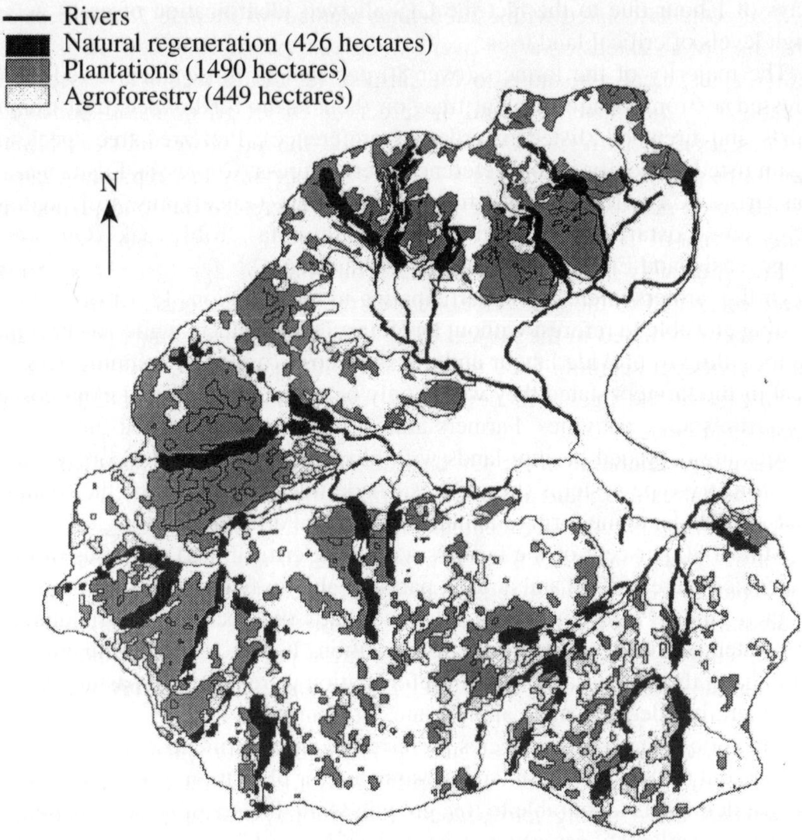

Figure 2.5 Proposed reforestation plan for the Purire River watershed

Estimating the costs of the reforestation plan

The final step in the planning of a reforestation plan for the upper Purire watershed involved determining the costs of establishing 1490 ha of plantations, 449 ha of agroforestry systems and 426 ha of river and stream buffers. Based on discussions with local farmers and community leaders, it was assumed that these reforestation plans should be organized and implemented through the local community-based agricultural cooperative which already had experience in promoting reforestation plans to the local farmers in the area.

A hypothetical yet realistic plan would involve hiring a full-time extension agent for a three-year period, the use of subsidized (low-cost) tree seedlings produced in local nurseries and the provision of technical assistance to farmers applying for reforestation subsidies from the national government.

The costs associated with establishing plantations include 75 per cent of the costs of an extension agent over a three-year period ($56 250) in order to provide technical assistance to the farmers regarding technical aspects of establishing the plantations and the procurement of government-sponsored reforestation incentives. Other costs include the provision of trees (1390 seedlings per hectare, assuming a 25 per cent mortality and replanting rate). At 10 cents per seedling, this cost $165 700. Therefore the total plantation forestry cost of the project is $221 950.

Specific costs associated with the promotion of agroforestry systems were estimated to include 25 per cent of the three-year costs of the extension agent assigned to the project ($18 750), and 150 seedlings per hectare at 15 cents per seedling for a total of $10 100. In addition, $1000 worth of fertilizer for the trees will be needed for a total cost of $29 850 over a three-year period.

Natural regeneration of the 426 ha of identified critical area lands buffering rivers and streams should not cost the project any additional funds. A requirement for farmers receiving technical assistance and free seedlings under the agroforestry or plantation forestry components of the project should be that these buffer areas be left to regenerate naturally over time.

In summary, the total project costs of establishing 1490 ha of plantations, establishing agroforestry and soil conservation practices on 449 ha, and allowing the regeneration of 426 ha of river and stream buffers would be approximately $252 000.

This case study has demonstrated the feasibility of the plan and estimated the costs of an efficient and feasible reforestation plan for a degraded tropical watershed while taking into account both biophysical and socioeconomic factors. Such an interdisciplinary and comprehensive approach required the use of GIS technology and a concerted multidisciplinary work effort.

Case Study 2: Resource Planning in the Pristine Pacuare Watershed, Costa Rica

The Pacuare is a multiple-use tropical river located on the Atlantic slope of Costa Rica. The upper Pacuare watershed is approximately 58 km long and 11 km wide, with an overall area of 64 000 ha, and its headwaters begin in the Talamanca mountains and flow north onto the Atlantic coastal lowlands of the Caribbean. With minimal logging and agricultural interventions in the watershed, the Pacuare remains one of the most pristine and scenic rivers in Costa Rica, with clean water and a great diversity of native flora and fauna. The Pacuare has also become one of the premier whitewater rafting rivers in the world, and it has also been identified as a proposed site for a 400-megawatt electric dam.

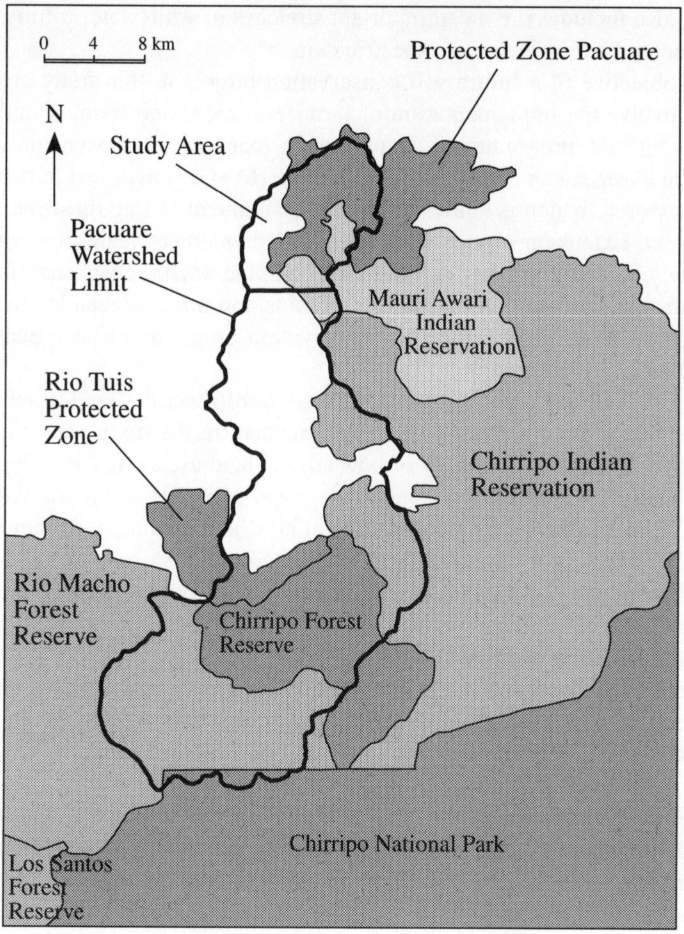

Figure 2.6 The Pacuare River watershed in Costa Rica

As shown by Figure 2.6, a large part of the watershed is located near or within several protected areas and an Indian reservation which, along with the upper and southern parts of the watershed, are somewhat inaccessible and are dominated by natural forests. The northeastern or lower part of the watershed has relatively good road access and is in close proximity to several small towns. It also has the highest level of privately owned farms and agricultural land uses in the watershed, and will likely be the focus of any future soil conservation programmes in the watershed. This lower section was therefore chosen as a pilot study area. It contains about 9000 ha of primary forests, cattle ranches and small-scale hillside farming activities. As mentioned before, it also includes the most important stretches of whitewater rafting along the river, and a proposed hydroelectric dam.

The objective of a future soil conservation project in this study area will likely involve the implementation of farm-level extension training and technology transfer programmes. At present, a team of four extension agents from the Costa Rican Ministry of Agriculture (MAG) is assigned part-time to the study area, which is equivalent to the assignment of one full-time extension agent. Extension service work is equally divided between improving the efficiency of cattle grazing activities and helping small subsistence farmers diversify their production of cilantro, orchids and other vegetables for local and export sales. No specific soil conservation programmes have ever been implemented in this area.

Critical land use areas in the watershed were identified and a set of resource management practices was recommended for the study area (Shultz *et al*., 1998). Below is a summary of how critical land use areas were identified, farms spatially georeferenced, and farm ownership and production data collected. This is followed by a description of how land use and degradation data were classified by alternative farm sizes in order to formulate a resource management plan for the area.

The identification of critical land use areas

Land uses and degraded lands in the study area were classified using GIS-based spatial overlays of the following biophysical data themes: geology and soils, elevation, vegetative cover, land use and land use capacity. Degraded lands were classified into four distinct categories: high, medium, low and none. Only 10 per cent of the lands within the study area were classified as suitable for intensive agricultural uses (annual crops and grazing) versus 44 per cent for perennial crops (coffee, ornamentals and plantation forests) and forests, and 30 per cent for long-term natural forests. Finally, 16 per cent was classified as protected, with no specific human interventions or activities recommended. As shown by Figure 2.7, the dominant land use in the study area is primary natural forest (82 per cent of the total), followed by pasture for cattle grazing (11 per cent), while secondary forests, plantations and cultivated areas are less than 7 per cent of the total land uses.

Various combinations of land 'overuse' were spatially overlaid with estimated levels of erosion in order to classify levels and locations of degraded lands. The results indicated that only about 1964 ha or 22 per cent of the study area contain lands that are degraded. This includes 682 ha (7 per cent) of high-level degradation, 238 ha (3 per cent) of medium-level degradation and 1080 hectares (12 per cent) of low-level degradation.

As expected, most of the degraded lands are located in the northeastem side of the pilot area of the watershed where the highest concentration of

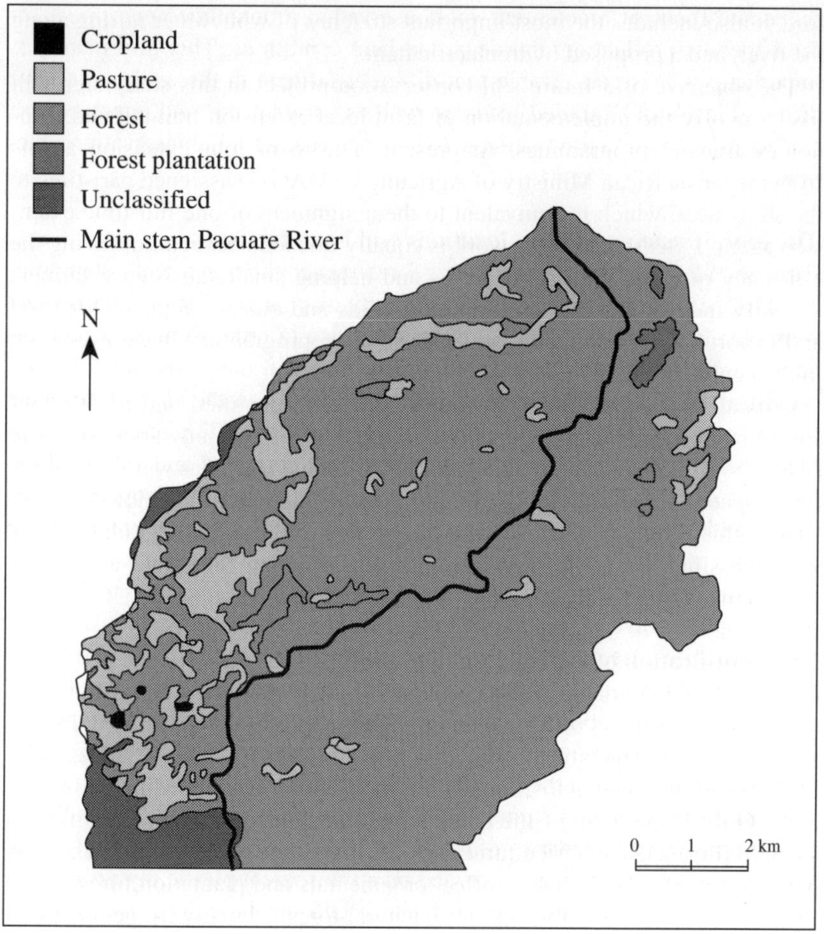

Figure 2.7 Land uses in the Lower Pacuare River watershed

towns, roads, private farm holdings, and deforested pasture and croplands are located. However, there are also pockets of lands without any degradation within this generally heavily degraded area, as there are pockets of highly degraded lands within the generally non-degraded parts of the study area. Such land degradation data are very useful for the general planning of a soil conservation programme, but at the same time they are limited because little is known about the number and size of actual farms with different levels of land degradation in the study area. Similarly, little information is available about the relationships between farm ownership, production patterns and land use degradation. This means that it is difficult to determine the intensity, scope and focus of a potential soil conservation programme appropriate for farm-specific production and degraded land conditions. This information is important, as it is conceivable that different farmers in the study area have different levels of degraded lands, as well as varied needs and interests with respect to alternative production activities and related soil conservation practices.

The georeferencing of farm locations and boundaries

Farm boundaries and locations were georeferenced through a multi-step process. First, cadastral land ownership property maps and a list of farm owners were obtained from the national cadastral office in order to make a preliminary identification and contact with farm owners in the study area. A team consisting of a local farmer, neighbours (especially in the case of absentee owners or managers), a local extension agent and a CATIE researcher then made farm visits. Together this team identified, corrected and validated the locations of farm property boundaries while using cadastral property maps, topographic maps and an enlarged black-and-white air photograph of the study area as references. Visual references were made between known farm boundaries and a variety of geographic features visible on the maps and air photograph, including roads and trails, fields and clearings, housing and other structures, tree stands, rivers and streams, fences and field and forest boundaries. In the cases where difficulties in identifying farm boundaries occurred, a visualization exercise was made from a high-elevation strategic location with an overview of the farm area in question. And, in some cases, GPS was used to obtain the geographical coordinates of property boundaries.

From these combined strategies, it was possible to identify, and validate, the farm property boundaries of 41 of the 50 known farms in the study area. These farm boundaries were recorded on a transparent plastic mylar overlaid on the air photo, and later digitized into a GIS database. The use of an air photo, while it contained some distortion, provided reasonably consistent and accurate indications of farm boundaries. Finally, it is interesting to note that over 75 per cent of cadastral maps and about 30 per cent of the property owner classifications were not accurate.

Farms were then classified by size, with small farms being defined as being less than 15 hectares, medium farms between 16 and 75 hectares, and large farms greater than 75 hectares. The eight large farms averaged 358 ha, while medium farms averaged 33 ha and small farms 7.5 ha.

The combined total labour requirements associated with the georeferencing of farm boundaries were approximately three months of an individual person's full-time labour. Furthermore, all of the required computer equipment and related materials for these farm-level georeferencing activities had a cost of less than $2000 (not including fixed or start-up costs associated with a GIS laboratory). From this experience it was concluded that the total time and costs of georeferencing property farm boundaries in a typical, medium-sized pristine Central American watershed are relatively low, especially when considering the many potential uses of the resulting spatially referenced farm-level data.

The integration of socioeconomic and biophysical data

Ownership, production and farm value data were collected through personal surveys of the farmers in the pilot study area and classified and analysed by farm size. It is important to note that all of these farm-specific data are spatially referenced, and that they were collected at the same time as the farm-level property mapping exercises. The collected data were somewhat limited in scope and detail because many of the surveyed farmers have low levels of literacy or limited record-keeping skills and time available to participate in such surveys.

Ownership data were used to determine whether farms were occupied by owners, renters or squatters, while production data were used to confirm and/or update the farm-level land use estimates. General land use data from the earlier GIS-based study (and the spatial overlay with farm boundaries) were almost always found to be consistent with farmers' perceptions of their own land uses. However, in some cases the farmers were able to specify more accurate estimates of the breakdown of annual versus perennial crops on their lands. They were also able to supply information related to grazing densities, irrigation systems and the use of soil conservation methods, not all of which was available from the earlier GIS-based land use study.

These ownership and production data were then integrated with biophysical based land use and land degradation data (estimated in the previous study in the area), through two GIS spatial overlays. First, the coverage of farm property boundaries was spatially overlaid with the land use coverage, followed by a spatial overlay of the land degradation coverage. A part of this integrated GIS database is illustrated in Figure 2.8, which is a map showing both farm property boundaries and levels of medium and high land use degradation throughout the study area. These data classified by small, medium and large farm sizes are summarized in Table 2.1. Also included are the results of bivariate (chi-square) statistical tests of differences in land use, degradation and farm characteristics by farm size classifications.

Regarding observed farm-level land uses and production activities in the study area, the following information relevant to the design and implementation of a soil conservation programme was noted. Small farms have more diversified and intensive agricultural activities, are more deforested, have higher grazing densities, and a higher percentage of annual and perennial crops than do medium or large-size farms. Also large farms have very few

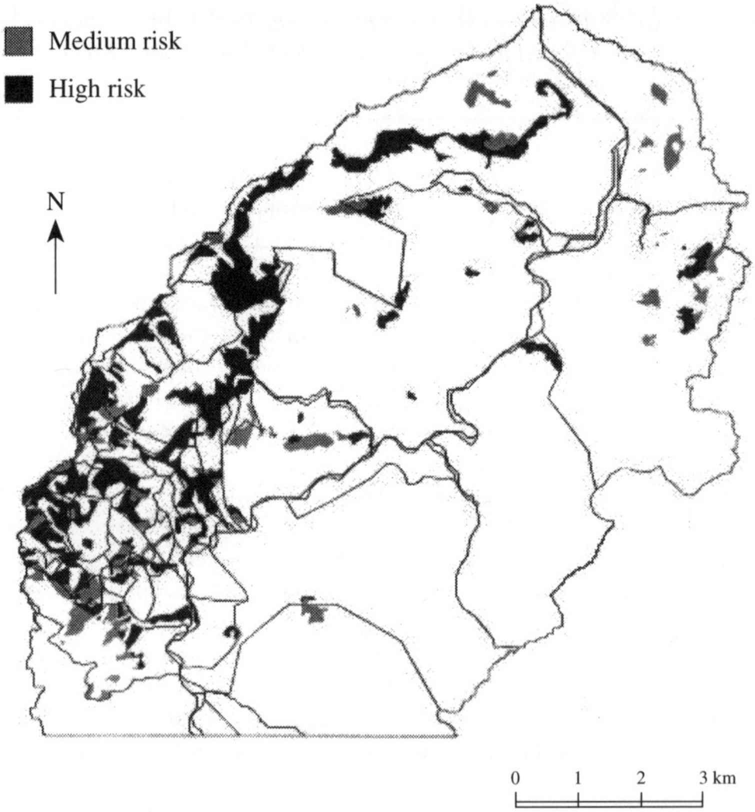

Figure 2.8 Farms with medium and high levels of degradation risk, based on biophysical and socioeconomic data, in the Lower Pacuare watershed

cultivated crops and significantly larger percentages of their lands dedicated to natural and plantation forests than do medium and small farms. Finally, medium-sized farms have moderate levels of annual and perennial crops and a higher proportion of their total land area in secondary growth (brush). Finally, almost 13 per cent of medium-sized farms were found to be using some form of soil conservation technologies, while small and large farms were not found to be using any soil conservation practices.

These integrated farm size, land use and production data would seem to indicate that the most appropriate soil conservation technologies for small farms in the study area should include two distinct approaches: first, to reduce soil erosion associated with annual and perennial crops such as con-

Table 2.1 Spatially integrated ownership, production, land use and degradation data by farm size in the lower Pacuare River watershed

	Small farms (<15ha)	Medium farms (16–75ha)	Large farms (>75ha)
Number	18	15	8
Farm size (ha)	7.5	33[a]	358[b,c]
Self-owned farms	89%	87%	100%
% land of high degradation	8	3[a]	2[b]
% land of medium degradation	4	1[a]	2[b]
% land of low degradation	3	5[a]	7[b]
Weighted measure of degradation (3*high % + 2*med % + 1*low %)	35	16[a]	17[b]
Pastures	44%	59%	52%
Annual crops	13%	3%[a]	0%[b]
Perennial crops	9%	3%[a]	4%[b]
Primary forests	9%	8%	25%
Secondary forests	1%	1%	2%
Plantation forests	4%	3%[a]	12%[b,c]
Secondary growth	20%	23%	5%[b,c]
Grazing density of farms with cattle and horses (animals/ha)	2.7 (n=7)	1.1[a] (n=7)	1.2[b] (n=5)
Use of soil conservation	0%	13%[a]	0%[c]
% farms with water sources	89	93	88
Mean houses per farm	1.1	1.4	3.1[c]

Notes:

[a] Significant difference between small and medium farms at the 95 per cent confidence level.
[b] Significant difference between small and large farms at the 95 per cent confidence level.
[c] Significant difference between medium and large farms at the 95 per cent confidence level.

tour planting, live and dead barriers, drainage ditches and mulching systems, and, second, to promote more sustainable (less intensive) cattle grazing management practices. In contrast, soil conservation programmes for medium-sized farms should be more focused on secondary forestry management options and possibly the development of plantation forestry systems. Also the focus of soil conservation programmes for large farms should be on natural forestry management options or, in other words, strategies that will ensure that farmers maximize the long-term profit potential of their forests, either through selected and sustainable logging or through ecotourism activities.

These recommendations for alternative soil conservation strategies for farms of different sizes are somewhat generalized, in that they are based on average data across farm types. However, they are still a major improvement on the previous situation, where local extension agents and natural resource planners had only a general and fragmented knowledge of the relationships between land degradation and farm types. It is also the case that the spatially integrated database of farm boundaries, land uses, ownership patterns and production activities can be analysed at the individual farm level among the 41 farms in the study area. Therefore, if desired, farm-specific data can be readily assessed and used to determine the most appropriate soil conservation methods tailored to the specific needs of the people who live and work on individual farms.

The integrated relationship between land use, land degradation and property boundaries is also very useful for prioritizing farms (either individually or by groups of similar farms) that need to be changed by a future soil conservation plan in the study area. For example, it can be seen from Table 2.1, that, on average, 8 per cent of small-sized farms are highly degraded, while medium and large-sized farms are on average only moderately degraded.

To obtain more site-specific degradation data, a relative indicator of total farm-level degradation was calculated for each farm. High degradation was given a relative weight three times as large as the value of low degradation, while medium degradation was weighted as twice as important as low degradation. This simplified weighting strategy was based on the straightforward assumption that a hectare of highly degraded land is more degraded than a hectare of medium degraded land, which in turn is more degraded than a hectare of low degradation. These degradation indices are a relative measure of the extent of total degradation. As expected, small farms have significantly higher levels of total land degradation than do medium and large farms.

By evaluating the spatial relationship between land degradation and farm sizes, it is possible to rank and hence prioritize the type of farms most in need of soil conservation technologies. Again, this prioritization exercise can be done either at the individual (farm) or at the group (large, medium or small-sized farms) level of analysis. For example, it is clear that, as a group, small farms within the study area have the highest levels of land degradation. In the event of a future soil conservation programme in the area with limited available financial resources, these small farms should be given priority over medium and large-sized farms. And, in this case, the prioritization of individual farms based on their relative levels of total land showed results somewhat consistent with those at the group level of analysis, as eight of the 10 farms with highest levels of total degradation were small farms.

CONCLUSIONS FROM THE CASE STUDIES

The two case studies presented in this chapter have demonstrated the benefits of using GIS technology to integrate biophysical and socioeconomic data in order to plan various sustainable resource management activities. In the first case study, a GIS was used to identify critical land use areas with information concerning the preferences and desires of local farmers in order to produce the details and costs of a reforestation programme. In the second case study a GIS was also used to identify critical land uses which were linked to socioeconomic data through the mapping of individual farm boundaries within the study area. This integration of data was then shown to be extremely useful for resource planning activities through the classification of different activities (soil conservation, reforestation or natural forest management) deemed most appropriate to the specific needs of farms based on their size and characteristics. It also permitted the prioritized ranking of farms most in need of soil conservation based on their relative levels of land degradation.

It is hoped that other natural resource professionals will use these or other relatively simple and available methodologies to plan comprehensively the sustainable management of resources in other tropical areas of the world. In summary, the major advantages of using GIS technology are the efficient identification and prioritization of critical land use areas within a watershed or study area while taking into account both biophysical and socioeconomic data. Such integration is considered essential for planning and evaluating specific sustainable resource management plans over space and time.

It should also be noted that, in spite of limited funding and many technical and data constraints in developing countries, the methodologies demonstrated in this chapter for integrating biophysical and socioeconomic data are technically feasible at a relatively low cost. It is therefore hoped that more natural resource management practitioners working in Central America and other developing and tropical countries will begin actively to use these methodologies for more effective planning, and to implement soil conservation and other natural resource management programmes.

This should become much easier in the coming years as more biophysical and socioeconomic data in these regions become spatially referenced, and as GPS technologies continue to improve and become more affordable. However, to facilitate this process, national and local governments should spatially reference and better maintain cadastral surveys, and make this information compatible with national GIS databases that are currently being developed. It is also hoped that national census programmes in Central America will begin to collect and report their data at more detailed levels of census geography. Finally, it is recommended that social scientists working in Central America who collect such data should make greater efforts to identify the geographical locations of the farms from which they collect data.

REFERENCES

Adinarayana, J., J. Flach and W. Collins (1994), 'Mapping Land Use Patterns in a River Catchment using Geographical Information Systems', *Journal of Environmental Management*, 42, 55–61.

Centro Cientifico Tropical (Tropical Science Center) (1994), 'Report on Land Use Maps and Forested Lands in Costa Rica', San José, Costa Rica: CCT.

Current, D., E. Lutz and S. Scherr (1995), 'The cost & benefits of agroforestry to farmers', *The World Bank Research Observer*, 10(2), 51–180.

Jimenez, F.O. and J.M. Faustino (1996), *Summaries of Msc. Thesis Projects in the Watershed Management Area of CATIE (1963–1996)*, Watershed Management Technical Publication Series # 1, CATIE, Turrialba, Costa Rica.

Lutz, E., S. Pagiola and C. Reiche (1995), 'The costs and benefits of soil conservation: the farmers' viewpoint', *The World Bank Research Observer*, 9(2).

Maguire, D.J. (1991), 'An overview and definition of GIS', in D.J. Maguire, M. Goodchild and R. Rhind (eds), *Geographic Information Systems: Principles and Applications*, London: Longman.

Medina, J.M, S. Shultz and S. Velasquez (1997), 'Planificacion de la reforestacion de una Cuenca Degradada utilizando una Sistema Informacion Geografica', *Agroforesteria en Las Américas* (Agroforestry in the Americas), June.

Shultz, S.. F. Saenz and G. Hyman (1998), 'Linking People to Watershed and Protected Area Planning with a GIS: A Case Study of the Rio Pacuare, Costa Rica', *Society and Natural Resources*, 11(6), 663–75.

Theocharoppolos, S.P., D.A. Davidson, J.N. McArthrur and F. Tsouloucha (1995), 'GIS as an aid to soil surveys and land evaluation in Greece', *Journal of Soil and Water Conservation*, March–April.

3. Carbon storage in soils and vegetation among forested ecosystem types in northern Ontario

Mark Johnston and Peter Uhlig

INTRODUCTION

Carbon (C) storage has received increased attention as the reality of global warming is recognized (Houghton *et al.*, 1996). Forests in northern latitudes are thought to play an important role in global C storage (Tans *et al.*, 1990; Fan *et al.*, 1998) and may be disproportionately affected as global warming intensifies (Pastor and Post, 1988; Kirschbaum and Fischlin, 1996). The boreal forest is the world's largest depository of C (Dixon *et al.*, 1994), yet estimates of C storage for this biome are generally lacking. Botkin and Simpson (1990) and Kurz *et al.* (1992) presented C storage data for boreal forest regions, but these were for large, continent-wide areas. Some site-specific estimates are available (for example, Halliwell *et al.*, 1995; Nalder and Merriam, 1995), but knowledge of how C is distributed among ecosystem types within the boreal forest is generally poor.

We present estimates of C storage in soils and vegetation among 11 ecosystem types in a 180 000 km^2 area of northern Ontario that was intensively sampled during the development of the Ontario Forest Ecosystem Classification (FEC) system (Sims *et al.*, 1997). We compare these estimates with estimates of vegetative C storage for the same area taken from the most recent Forest Resource Inventory data for the Province of Ontario (Callaghan, 1993).

METHODS

Study Area and Sample Design

The area sampled is shown in Figure 3.1. It spans most of the northwest portion of Ontario, extending from the Manitoba border in the west to the

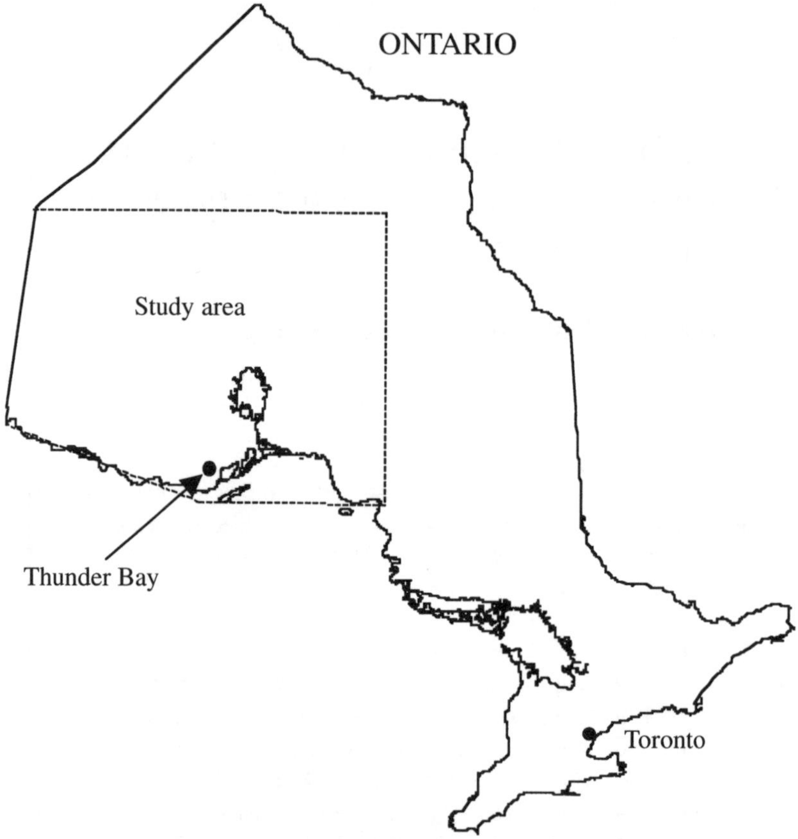

Figure 3.1 *Location of forest ecosystem classification study area in northern Ontario*

White River in the east. The United States border and Lake Superior form the southern boundary, and the northern extent is roughly coincident with the southern boundary of the Hudson's Bay Lowlands (Anonymous, 1974). This area was sampled in 1985–7 during the development of the Ontario FEC (Sims and Uhlig, 1992). Approximately 2500 plots were established across the area, with the goal of sampling mature (more than 50 years old) stands of commercial timber species.

The 38 original vegetation types identified by the FEC were grouped into 11 FEC groups, for which we characterized C storage in soil and vegetation. The vegetation types and FEC groups are given in Table 3.1. At each sample location, a 10 by 10 metre plot was established. The vegetation in each stratum was identified as to species and measured for percentage cover, and

Table 3.1 FEC groups and vegetation types (v-types) identified by the Ontario forest ecosystem classification

FEC groups	V-types
Miscellaneous hardwoods	Balsam poplar (V1), black ash (V2), other hardwoods (V3)
Birch hardwood and mixedwood	White birch hardwood or mixedwood (V4)
Aspen hardwood and mixedwood	Aspen hardwood (V5), other V-types with aspen dominant in overstorey (V6–V11)
Red or white pine conifer or mixedwood	White pine mixedwood (V12), red pine mixedwood (V13), white pine conifer (V26), red pine conifer (V27)
Balsam fir/white spruce mixedwood	Balsam fir mixedwood (V14), white spruce mixedwood (V15), balsam fir/white spruce mixedwood – feathermoss (V16)
Jack pine – shrub rich	Jack pine mixedwood – shrub rich (V17), jack pine – low shrub (V28)
Jack pine – feathermoss	Jack pine mixedwood – feathermoss (V18), jack pine – ericaceous shrub – feathermoss (V29)
Black spruce/jack pine – feathermoss	Black spruce mixedwood – herb rich (V19), black spruce mixedwood – feathermoss (V20) black spruce/jack pine – tall shrub – feathermoss (V31), jack pine/black spruce – ericaceous shrub – feathermoss (V32), black spruce – feathermoss (V33)
Black spruce – wet organic	Lowland sites with organic or mixed organic/mineral soils with overstorey dominated by cedar (V22), tamarack (V23) or black spruce (V34–V37)
Jack pine/black spruce – blueberry – lichen	Jack pine/black spruce – blueberry – lichen (V30)
Black spruce – leatherleaf – sphagnum	

Source: Sims *et al*. (1997).

trees were measured for diameter at breast height (DBH) and height. In each plot a soil pit was excavated and the soil profile was described. Soils were sampled by horizon and analysed following the procedures of Kalra and Maynard (1991). Organic matter content of soil and forest floor was estimated by weight loss following combustion of samples at 375°C for 16 hours (ibid.).

Data Analysis

C storage in the tree stratum was estimated by determining biomass for each tree on a plot from height and diameter measurements using species-specific whole-tree estimation equations for Ontario (Alemdag, 1983; 1984). Biomass was converted to C by assuming a C content of 50 per cent (Turner *et al.*, 1995) and was summed across plots and converted to kg/m^2. Understorey biomass could not be calculated due to lack of appropriate data for the application of biomass equations to herbs and shrubs. Instead, we used understorey biomass values reported by Ohmann and Grigal (1985) for a series of ecosystem types in northern Minnesota similar to those identified by the FEC. In any case, understorey biomass is rarely more than 2–3 per cent of total above-ground biomass in boreal forest stands (Johnston *et al.*, 1996), so the lack of specific understorey biomass data was not considered a significant drawback in this study. Understorey biomass values from Ohmann and Grigal (1985) were converted to C content using a conversion factor of 50 per cent.

Soil C content (percentage by weight) was estimated from loss-on-ignition (LOI) values using a relationship for the northern Great Lakes region of the United States (David, 1988). Bulk density was determined from LOI using a BD–LOI relationship from the same area (Grigal *et al.*, 1989) and horizon thickness was taken from the plot data. Soil C storage was calculated by multiplying horizon thickness, bulk density and C content and converting to kg/m^2. We determined the standard error of the mean (SE) for all estimates and propagated the SE among calculations using methods given in Freese (1962).

The FEC data were collected as part of an ecosystem classification effort in Ontario. However, many forest estates do not have data of this level of detail available for calculation of C storage. Traditional sources of these data are forest inventory data, used for determining commercially useable timber volumes and growth rates (for example, Brown and Lugo, 1984). We compared the FEC data with forest inventory data for the study area in order to determine whether ecosystem survey data would yield estimates of C storage similar to those derived from inventory data. The inventory data were taken from the most recent Forest Resource Inventory (FRI) for Ontario (Callaghan, 1993). We aggregated the FRI data into species groups that corresponded as closely as possible to those identified in the FEC groups. C storage and density was determined by converting timber volumes to whole-tree biomass using relationships given in Alemdag (1983; 1984).

RESULTS

Mean C storage in vegetation for each FEC group is shown in Table 3.2. Vegetative C density was greatest in the red pine/white pine type and several of the hardwood mixed wood types, while that of jack pine/black spruce and black spruce–leatherleaf types was the lowest. C density in understorey vegetation was 1–3 per cent of total vegetative C in all groups except the jack pine/black spruce and black spruce–leatherleaf types, in which it comprised 7 per cent and 25 per cent, respectively.

C density in soil (O horizon and mineral soil) is also shown in Table 3.2. Black spruce on wet organic sites had the greatest C density in the O horizon. The lowest amounts were found in jack pine/black spruce and jack pine–feathermoss types and in birch mixed woods. Data on cumulative C density with depth were combined among all FEC groups and are presented in Figure 3.2. Even with the wide variation in C density values shown above, there was a relatively strong relationship between density and depth among FEC groups.

Total C density for each FEC group is shown in Table 3.2. The highest densities are in the red/white pine, black spruce–wet organic and miscellaneous hardwoods types. Each group varied in the ecosystem compartment in which the C density was greatest: vegetative C was greatest in the red/white pine type, O horizon C greatest in the black spruce–wet organic type and miscellaneous hardwoods had high densities in both mineral soil and vegetation.

The extent to which ecosystem compartments are affected by disturbance and management activities varies among ecosystem types. If the impacts of harvesting are of concern with regard to C release, the ratio of vegetative to soil C can indicate the potential impacts of harvesting among the ecosystem types. Figure 3.3 shows this ratio for the FEC groups. The red/white pine type had the greatest proportion of C in vegetation; in fact it was the only FEC group in which above-ground C was greater than that in soil.

The comparison of C density derived from the FEC data with that from the FRI data is shown in Figure 3.4. The FRI data yielded estimates that were from 28 per cent to 76 per cent of those from the FEC data. The greatest differences were in the red/white pine type (28 per cent of FEC values) and balsam fir/white spruce mixed wood type (32 per cent of FEC values).

Table 3.2 Carbon density (mean and SE) for soil and vegetative components among 11 ecosystem types in northwestern Ontario (mineral soil carbon up to 100cm) depth

FEC group	Tree C	SE	Understorey	SE	O horizon (kg/m^2)	SE	Mineral soil	SE	Total C	SE
Miscellaneous hardwoods	12.0	1.1	0.19	0.02	10.7	8.1	14.9	2.6	37.8	13.6
Birch mixedwood	4.5	0.5	0.12	0.03	4.0	0.5	13.6	6.3	22.2	20.0
Aspen mixedwood	9.6	0.5	0.21	0.03	9.1	7.0	10.8	1.1	29.7	10.0
Red or white pine mixedwood	17.9	1.3	0.15	0.05	5.2	1.5	11.1	1.3	34.4	4.3
Balsam fir/white spruce mixedwood	10.1	0.7	0.12	0.03	8.1	3.4	13.0	3.3	31.3	14.9
Jack pine – shrub rich	7.8	0.6	0.15	0.03	7.0	1.5	7.3	0.8	22.3	3.5
Jack pine – feathermoss	3.9	0.3	0.15	0.03	3.6	0.4	6.4	1.5	14.1	4.5
Black spruce/jack pine – feathermoss	6.4	0.3	0.15	0.03	10.1	8.9	10.3	1.7	27.0	11.9
Black spruce – wet organic	5.3	0.4	0.15	0.03	12.8	7.1	13.5	2.7	31.8	13.9
Jack pine/black spruce – blueberry – lichen	2.1	0.2	0.15	0.03	3.8	0.5	13.5	3.2	19.5	5.2
Black spruce – leatherleaf – sphagnum	0.6	0.2	0.15	0.03	0.0	0.0	18.5	14.8	19.3	11.9
Mean	7.3	0.7	0.20	0.03	6.8	5.6	12.1	4.5	26.3	12.4

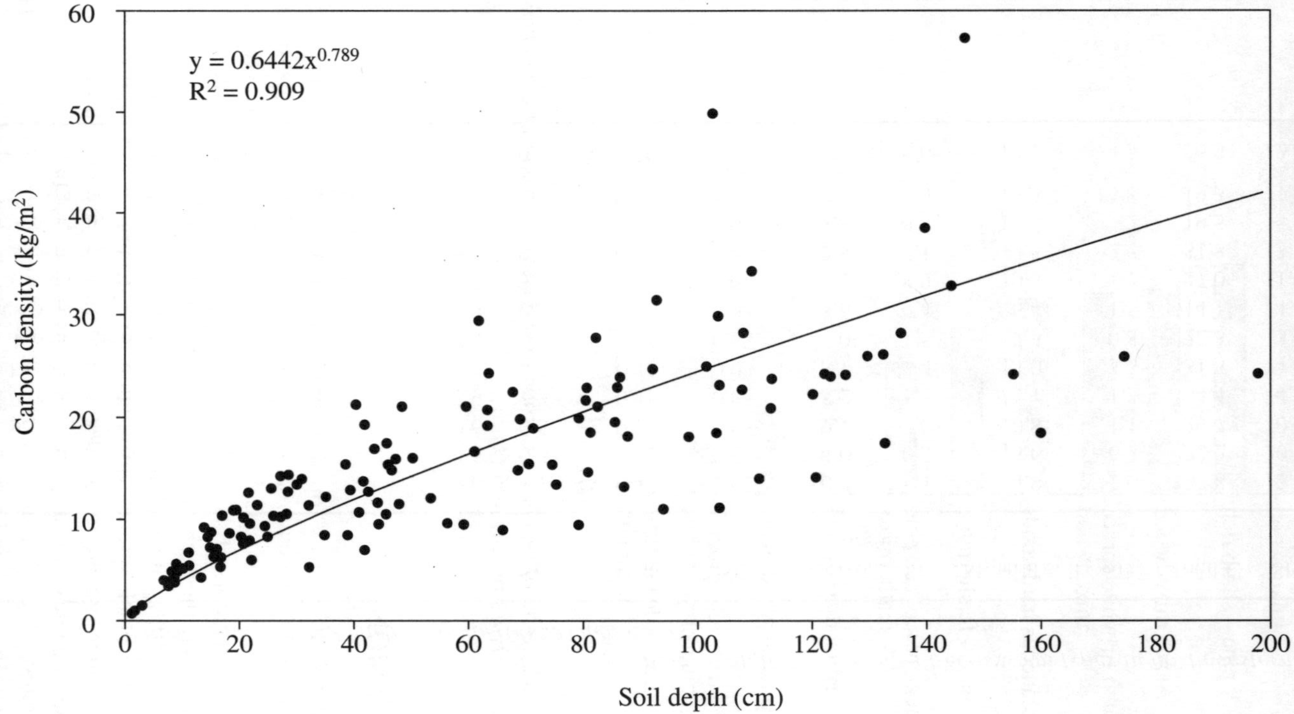

Figure 3.2 Carbon density with soil depth for 11 ecosystem types combined in northern Ontario

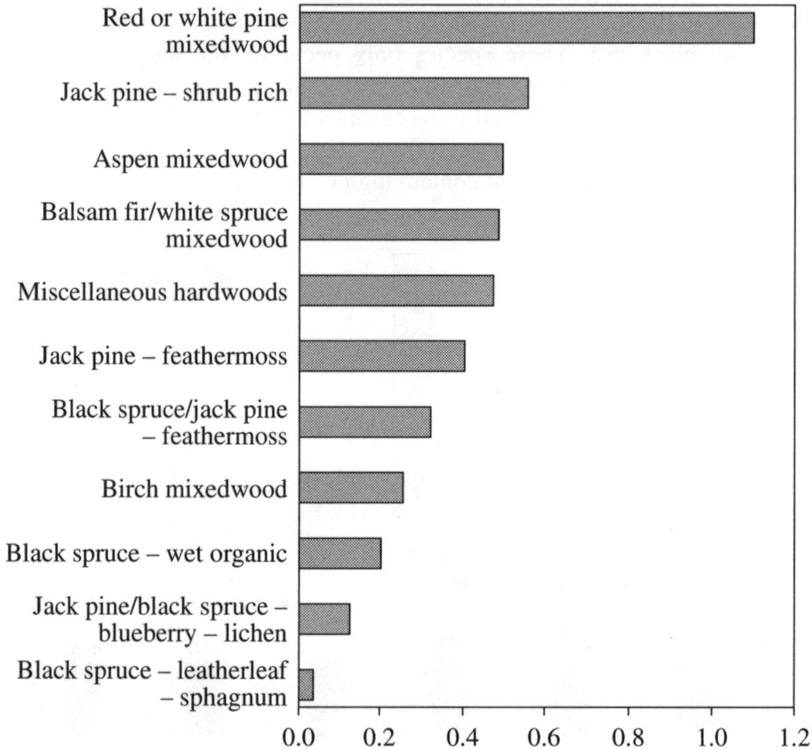

Figure 3.3 *Ratio of vegetative carbon to soil carbon among 11 ecosystem types in northern Ontario*

DISCUSSION

Most of the differences found among ecosystem types can be explained with reference to the environmental conditions in which each type occurs. High levels of vegetative C density in the red/white pine type are a result of the long-lived nature of these species and high levels of biomass accumulation. However, they tend to occur on relatively dry coarse-textured soils (Sims *et al.*, 1997) in which organic matter does not tend to accumulate. In contrast, the black spruce–wet organic type occurs on wet, acid lowland areas in which organic matter decomposition rates are low. In these environments, organic matter accumulates on the soil surface but rates of biomass accumulation through vegetative growth are low (Grigal *et al.*, 1985). The miscellaneous hardwood type is a broad category that includes many species found at the northern limit of their ranges, including red maple, yellow birch and black ash. These species only occur on the most favourable microsites in northwestern Ontario, typically moist, rich sites in sheltered locations (for example, alluvial terraces and south-facing slopes with deep soils; see Sims *et al.*, 1997). Accordingly, these sites are highly productive during the growing season and contain high C densities.

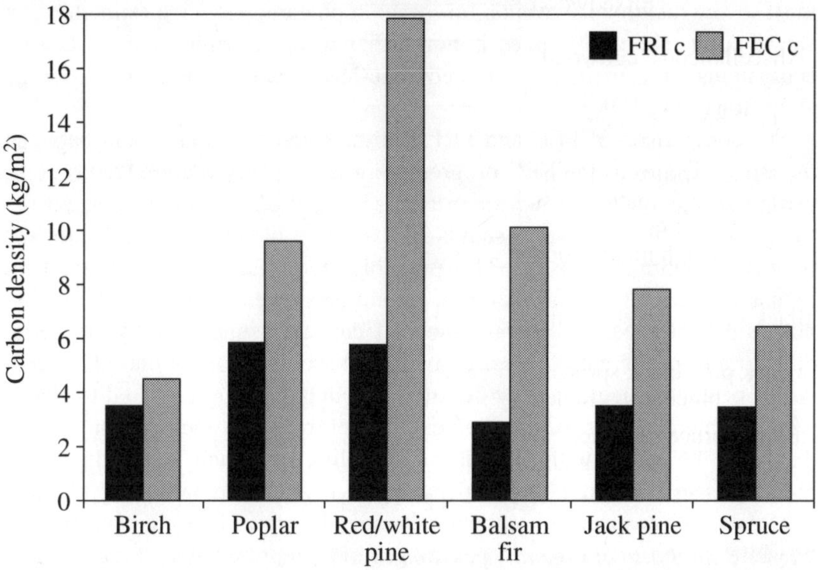

Figure 3.4 Comparison of carbon density derived from FEC and FRI data for forest land in northern Ontario

Potential effects of harvesting on C storage can be estimated by examining the patterns of C density among ecosystem compartments and their economic value. FEC types with high C density above-ground and high economic value, especially the red/white pine type, have the potential to release large amounts of C to the atmosphere, depending on the extent of harvesting and the ultimate fate of the biomass. In addition, many years will be required for this level of C density to become re-established due to their long lifespan (Harmon *et al.*, 1990). In contrast, FEC types such as black spruce–wet organic may have timber of value but the majority of C in this type is in the O horizon. The typical harvesting practice in these stands is to log in winter, thus having minimum impact on the frozen organic layer. Similarly, jack pine is a valuable species but frequently occurs on stony or sandy sites with low C densities.

Natural disturbance rates also vary among ecosystem types and have a strong influence on carbon storage at larger spatial and temporal scales. For example, well-drained upland ecosystem types probably experience fire more frequently than ecosystem types occurring on wet sites (Heinselman, 1981) and therefore accumulate less carbon between disturbance events. In addition, some species resist fire because of their large size and thick bark (for example, red pine types) and will continue to store carbon after all but the most severe disturbance events. Information on carbon storage estimated at a single point in time or space is not sufficient to determine the larger role ecosystems play in long-term landscape-level patterns of carbon storage (Johnston *et al.*, 1996).

The comparison of FEC and FRI data indicated little agreement between the two. The aim of the FEC programme was to survey mature stands comprising species of economic importance (Sims *et al.*, 1997). As a result the sampling was biased towards stands 50–100 years old that were fully stocked (that is, of optimum density). This probably represented the maximum stand biomass for each type and therefore should be viewed as an upper bound on potential C storage. In contrast, the FRI data were interpreted from aerial photographs of the entire forest estate in Ontario, and are not biased towards stands of any particular age or density (although they are restricted to stands with commercially important species). Therefore stands included in the FRI data may be older, with significant mortality, or young stands in which biomass is still accumulating. This is apparent when comparing the FRI and FEC data with respect to successional stage. The red/white pine and balsam fir/white spruce mixed wood types are generally representative of later stages of succession which would include significant mortality, and are the types for which there is the greatest disagreement between the FRI and FEC data. This may explain why the FRI data show much lower average C densities that do the FEC data.

Other work in this region has found generally lower values of C density. Grigal and Ohmann (1992) reported C densities in trees, mineral soil and O horizon for northern Minnesota, Wisconsin and Michigan. For trees, above- and below-ground C density ranged from 7 to 10 kg/m^2, with the highest values in forest types dominated by sugar maple, balsam fir and aspen. Their values include both above- and below-ground biomass, so values comparable to the present study would be even lower. Grigal and Ohmann's (1992) study was based on data collected on forest inventory plots, so the differences may be due to successional stage or mortality, as discussed above. They found C densities of 1 to 3 kg/m^2 in the O horizon, with the greatest values in stands dominated by sugar maple and balsam fir, values 30 to 50 per cent of those found in this study. In general, decomposition rates may be lower in the more northerly latitudes of Ontario, thus resulting in greater rates of accumulation in the forest floor. In the mineral soil Grigal and Ohmann found C densities of 7 to 14 kg/m^2, with the highest values in forest types dominated by balsam fir, sugar maple and aspen, similar to the pattern found in biomass. These values are very similar to those found in this study, suggesting that soil carbon storage is less sensitive to differences in climate and mortality than are biomass or the O horizon (Kurz *et al.*, 1992).

Botkin and Simpson (1990) estimated biomass and above-ground C densities for several large regions within the boreal forest of North America. They found an average C density of 3.2 kg/m^2 for the region including northwestern Ontario, a value much lower than that found using the FEC data. However, this value is similar to those we found using the FRI data for Ontario (excluding the red/white pine type) and again are probably related to differences in the sampling design of FRI versus FEC data collection. Botkin and Simpson suggest that previously published values are much higher because of the bias towards undisturbed stands, a conclusion that also applies to the FEC data for Ontario. Forest ecosystem classification data provide an advantage in that they include ecosystem compartments not included in traditional forest inventory data, especially understorey and soil. However, users of these data must be aware of any biases inherent in sampling design.

ACKNOWLEDGMENTS

We thank Sean MacMurray of the Ontario Ministry of Natural Resources for providing the FEC data and for extensive discussions about the methodology used in the Ontario FEC programme.

REFERENCES

Alemdag, I.S. (1983), 'Mass equations and merchantability factors for Ontario softwoods', Canadian Forestry Service, Environment Canada, Information Report PI-X-23.

Alemdag, I.S. (1984), 'Total tree and merchantable stem biomass equations for Ontario hardwoods', Canadian Forestry Service, Environment Canada, Information Report PI-X-46.

Anonymous (1974), *Atlas of Canada*, 4th edn, Ottawa, Ontario: Canada Department of Energy and Mines.

Botkin, D.B. and L.G. Simpson (1990), 'Biomass of the North American boreal forest', *Biogeochemistry*, 9, 161–74.

Brown, S. and A. Lugo (1984), 'Biomass of tropical forests: a new estimate based on forest volumes', *Science*, 223, 1290–93.

Callaghan, B. (1993), *The timber resources of Ontario – 1993*, Ontario Ministry of Natural Resources, Toronto, Ontario: Queen's Printer.

David, M.B. (1988), 'Use of loss-on-ignition to assess soil organic carbon in forest soils', *Communications in Soil and Plant Analysis*, 19, 1593–9.

Dixon, R.K., S. Brown, R.A. Houghton, A.M. Solomon, M.C. Trexler and J. Wisniewski (1994), 'Carbon pools and flux of global forest ecosystems', *Science*, 263, 185–9.

Fan, S., M. Gloor, J. Mahlman, S. Pacala, J. Sarmiento, T. Takahashi and P. Tans (1998), 'A large terrestrial carbon sink in North America implied by atmospheric and oceanic carbon dioxide data and models', *Science*, 282, 442–6.

Freese, F. (1962), 'Elementary forest sampling', U.S. Department of Agriculture, Agricultural Handbook no. 232, Washington, DC.

Grigal, D.F., C.G. Buttleman and L.K. Kernik (1985), 'Biomass and productivity of the woody strata of forested bogs in northern Minnesota', *Canadian Journal of Botany*, 63, 2416–24.

Grigal, D.F. and L.F. Ohmann (1992), 'Carbon storage in upland forests of the Lake States', *Soil Science Society of America Journal*, 56, 935–43.

Grigal, D.F., S.L. Brovold, W.S. Nord and L.F. Ohmann (1989), 'Bulk density of surface soils and peat in the north central United States', *Canadian Journal of Soil Science*, 69, 895–900.

Halliwell, D.H., M.J. Apps and D.T. Price (1995), 'A survey of the forest site characteristics in a transect through the central Canadian boreal forest', *Water, Air and Soil Pollution*, 82, 257–70.

Harmon, M.E., W.K. Ferrell and J.F. Franklin (1990), 'Effects on carbon storage of conversion of old-growth forests to young forests', *Science*, 247, 699–702.

Heinselman, M.L. (1981), 'Fire intensity and frequency as factors in the distribution and structure of northern ecosystems', in H.A. Mooney, T.M. Bonnicksen, N.L. Christensen, J.E. Lotan and W.E. Reiners (technical coordinators), *Fire Regimes and Ecosystem Properties*, USDA Forest Service, General Technical Report WO-26.

Houghton, J.T., L.G.M. Filho, B.A. Callander, N. Harris, A. Kattenberg and K. Maskell (1996), 'Climate Change 1995: The Science of Climate Change', Intergovernmental Panel on Climate Change, Cambridge, UK: Cambridge University Press.

Johnston, M.H., P.S. Homann, J.K. Engstrom and D.F. Grigal (1996), 'Changes in ecosystem carbon storage over 40 years on an old-field/forest landscape in east-central Minnesota', *Forest Ecology and Management*, 83, 17–26.

Kalra, Y.P. and D.G. Maynard (1991), 'Methods manual for forest soil and plant analysis', Forestry Canada, Northern Forestry Centre, Information Report NOR X-319.

Kirschbaum, M. and A. Fischlin (1996), 'Climate change impacts on forests', in R. Watson, M. Zinyowera and R. Moss, R. (eds), *Climate Change 1995: Impacts, Adaptations and Mitigation of Climate Change*, Cambridge, UK: Cambridge University Press.

Kurz, W.A., M.J. Apps, T.M. Webb and P.J. McNamee (1992), 'The carbon budget of the Canadian forest sector: Phase 1', Forestry Canada, Information Report NOR-X-326.

Nalder, I.A. and H.G. Merriam (1995), 'Simulating carbon dynamics of the boreal forest in Pukaskwa National Park', *Water Air and Soil Pollution*, 82, 283–98.

Ohmann, L.F. and D.F. Grigal (1985), 'Biomass distribution of unmanaged upland forests in Minnesota', *Forest Ecology and Management*, 13, 205–22.

Pastor, J. and W.M. Post (1988), 'Response of northern forests to CO2-induced climate change', *Nature*, 334, 55–8.

Sims, R.A. and P. Uhlig (1992), 'The current status of forest site classification in Ontario', *Forestry Chronicle*, 68, 64–77.

Sims, R.A., W.D. Towill, K.A. Baldwin, P. Uhlig and G.M. Wickware (1997), 'Field Guide to the Forest Ecosystem Classification for Northwestern Ontario', Field Guide FG-03, Ontario Ministry of Natural Resources, Northwest Science and Technology, Thunder Bay, Ontario.

Tans, P.P., I.Y. Fung and T. Takahashi (1990), 'Observational constraints on the global atmospheric CO_2 budget', *Science*, 247, 1431–8.

Turner, D.P., G.J. Koerper, M.E. Harmon and J.J. Lee (1995), 'A carbon budget for forests of the conterminous United States', *Ecological Applications*, 5, 421–36.

PART II

The Economic Valuation of Forests: Boreal and Tropical Case Studies

4. The carbon cycle and the value of forests as a carbon sink: a boreal case study

Mohammed Dore and Mark Johnston

INTRODUCTION

In the theory of renewable and non-renewable natural resources, what is the appropriate concept of scarcity rent? How is this rent likely to change over time? A review of literature on this subject shows that a number of economists have divergent views. Brown and Field (1978) and Fisher (1979) argue that measures of scarcity suffer from serious conceptual shortcomings. If attention is confined to exhaustible resources, even here, what constitutes scarcity rent (or 'user cost' or 'royalty rate') and how it changes over time remains controversial. In the standard Hotelling model, it is argued that the scarcity rent rises *monotonically* over time at the social rate of discount. In contrast, both Heal (1976) and Hanson (1980), using very different cost functions, show that the scarcity rent must decline monotonically to zero, as the resource approaches exhaustion.

On the other hand, Solow and Wan (1976) analyse conditions under which resource extraction costs rise with the depletion of higher-grade deposits and extraction turns to lower and lower-grade ores. They argue that along the optimal path the shadow price of a resource rises at the real rate of interest, but that the difference between price and marginal extraction cost, which they call 'degradation cost', declines monotonically over time to zero.

The divergence of views is replicated in the empirical studies too. For instance, Barnett and Morse (1963), Barnett (1979) and Johnson *et al.* (1980) all find that unit extraction costs in real terms have declined, so that these resources are in some sense becoming less scarce. Nordhaus (1974) also found that real prices of 11 major minerals fell over the period 1900–1970. In contrast, Smith (1979), Slade (1982) and Hall and Hall (1984) all find that real prices of natural resources are rising.

If we turn to renewable resources such as fish and forestry, it is not at all clear whether scarcity rent is deemed appropriate at all. The forestry literature is mainly concerned with optimal harvesting, although a number of forest ecologists have argued that forests are fast 'disappearing', so that the remaining forests are becoming scarce (FAO, 1995). But as far as we are aware no attempt has been made to compute a shadow price or scarcity rent of forests in mitigating global warming; it is this which is a central objective of this chapter.

In an attempt to reconcile the divergent and conflicting approaches to scarcity rent of natural resources, Farzin (1992) presents a generalized model which shows that *in general* the scarcity rent is non-monotonic. He shows that rising scarcity rent of exhaustible resources is a special case. He also presents an important corollary, that when the discount rate is zero (due, say, to intergenerational equity considerations) then, contrary to the Hotelling rule, scarcity rent of an exhaustible resource always decline monotonically over time to zero, irrespective of the form of the extraction cost function.

This chapter is concerned with valuing forests for an important ecological function, namely the ability of forests (at least those in Canada) to sequester carbon and so mitigate global warming. It turns out that Canadian forests are a net 'sink' of carbon: they capture more carbon than they release to the atmosphere (see Appendix 1 for details). But quite apart from being a carbon sink, forests hold soil and prevent erosion; they purify water and play an important role in the water cycle; they help maintain biodiversity; they are hosts to entire ecosystems. However, all of these other ecological functions will be ignored in this chapter; the focus is entirely on forests as a carbon sink. An optimal control model (that maximizes value added) is used to derive a shadow price of forests from the optimal solution, and the value of forest is computed via econometric techniques, with value added as numeraire. From this the marginal social opportunity cost of forests per hectare is computed. It is argued that it is this opportunity cost that must enter any harvesting decision model, or any other social cost–benefit calculation involving the use of forests.

Our model is consistent with the results of Farzin (1992), mentioned above. The computed shadow price of forests obtained by normalizing the shadow price which respects the ecological constraint is shown to be non-monotonic in general, and a monotonic shadow price can be obtained as a special case. However, the special case (Model 1, Trend Analysis with Constant Uptake) is instructive on an introductory thought experiment. We then use trend analysis with variable uptake (Model 2) and an ARIMA model (Model 3) to determine the marginal social opportunity cost of a hectare of forest.

The chapter is organized as follows: in the first section, the assumptions of the model are stated; in the second section, the theoretical model is stated; and in the third section, the model is estimated using standard econometric techniques.

ASSUMPTIONS

1. Let a social planner maximize a social welfare functional *w*, where *w* is a functional of natural and human capital stock *x*(*t*), a set of control variables *u*(*t*), where *u*(*t*) are policies such as harvesting forests, investing in conservation or other policies that improve *x*, and time *t*. Hence let

$$W = \max_{u(t)} \int_0^T [V(x(t), u(t), t)]dt \tag{4.1}$$

which is quasi-concave in the arguments, and *V* stands for value added. The functional *W* embodies a Paretian judgment such that valued added is maximized, irrespective to whom it may accrue.

2. The vector *x*(*t*) is the state at time *t* of the natural and human-made capital stock, and the time path of *x* (that is $\dot{x}$) is governed by

$$\dot{x} = f(x(t), u(t), t) \tag{4.2}$$

3. Equation (4.2), the equation of motion, is convex. The policies *u*(*t*) modify future *x*(*t*). The first component of the vector *x*(*t*) is $x_1(t)$ which is the 'state of the forest'. As this is the only component of the vector of interest in this chapter, we can drop its subscript without any possibility of confusion.

4. The following initial and terminal conditions are given:

$$x(t_0) = x_0 \tag{4.3}$$

$$x(T) = x_T \tag{4.4}$$

where x_T is defined to be sustainable (Dore and Ward, 1995), beyond *T* to infinity.

5. The social value of value added is 'valued at par' only when there is no externality, that is when carbon emissions are just equal to the carbon sequestration by forests. Furthermore, the social value of value added declines proportionally with the excess of carbon emissions over carbon sequestration.

THE MODEL

Logical Implications of the Assumptions

The logical implication of assumptions 1 to 3 is that a dynamic optimal solution to equations (4.1), (4.2) and (4.3) exists, where the optimized value of V is W^*. This in turn implies that there exist costate variables $\gamma(t)$, which are positive numbers.

From the maximum principle, it follows[1] that:

$$\frac{\partial W^*}{\partial x^*} = \gamma^*(t), \quad \forall t_0 \le t \le T \tag{4.5}$$

As usually interpreted, the costate variable (to a convex problem) measures the sensitivity of the optimized value of the objective function to a slight relaxation in the constraint.

If we define the reciprocal of $\gamma(t)$ to be $\lambda(t)$, it follows that the latter are also positive numbers. Writing equation (4.5) in discrete time, we have:

$$\frac{\Delta W}{\Delta x} = \gamma(t) = \frac{1}{\lambda(t)} \tag{4.6}$$

or

$$\Delta x(t) = \lambda(t)\Delta W(t) \tag{4.7}$$

From assumption 1, ΔW is value added in each time period, and $\lambda(t)$ is the shadow price of forests.

It follows that as the right-hand side of equation (4.7) has the dimension of value (the 'value' of value added weighted by the shadow price), the left-hand side must also have a value dimension: the change in the 'state' of forests is the *flow value of the forests*, with value added as numeraire. That is, we are valuing forest services in terms of a conventional macroeconomic category, namely value added.[2] It is equation (4.7) that is estimated using econometric techniques.

Modelling Carbon Sequestration

It has been established by Kurz *et al.* (1992) that Canadian forests are a net sink for carbon because of the age and species distribution of Canadian forests.

Let $q(t)$ be carbon sequestered per hectare in MT (metric tonnes). Then $q(t)$ is given by:

$$q(t) = \int_1^{\kappa} \int_0^t b \cdot \sigma(\tau) d\tau \cdot d\kappa - \int_0^{\kappa} \int_0^t z \cdot \rho(\tau) d\tau \cdot d\kappa \qquad (4.8)$$

where

b = number of trees of each species per hectare,
κ = number of species of trees,
z = number of trees of each species per hectare lost due to fire, insect damage or other human activities,
$\rho(\tau)$ = probability of loss.

Therefore $q(t)$ is the net carbon sequestered per hectare in time period t. If we let the total area of forest at time t be $a(t)$, then total carbon sequestered (in MT) is

$$Q(t) = q(t) \cdot a(t) \qquad (4.9)$$

This completes the formal modelling of carbon sequestration, or 'uptake'. Details of the estimation of carbon sequestered are given in Appendix 1.

Modelling the Shadow Price

The reasons why the concept of scarcity rent is regarded controversial by Brown and Field (1978) is that, first, it is non-observable, and there is no standard method of numerical estimation. Second, the most appealing version of a shadow price is the costate variable that would be derived from dynamic optimization, as done in this chapter. As stated earlier, the costate variable is the marginal increase in value added due to a slight relaxation in the constraint, and its reciprocal is a shadow price of forests in terms of value added. But before a costate variable can be used empirically, strictly speaking, the dynamic optimization model of equations (4.1) to (4.5) must be solved numerically. This raises the serious issues of observability and controllability (Dore, 1995), which cast doubt as to whether the model can in fact be solved numerically.

The failure of observability and controllability occurs because the dynamic constraint given in equation (4.2) is not based on some invariant physical law, such as the laws of physics and engineering. Thus, while observability and controllability hold for firing rockets into space, they fail for most large-scale economic applications.

Third, even if it were possible to solve for the numeral values of the costate variable and its reciprocal $\lambda(t)$, the complementary slackness conditions mean that the costate variable is positive if and only if the constraint is fully binding. For our problem it means that the 'forests must be fully utilized' with no excess capacity. Thus the costate variable measures the sensitivity of the objective function due to a slight relaxation of the constraint only when the constraint is binding, but is otherwise zero. As noted by Brody (1970), in a different context, this does not yield a satisfactory index of relative scarcity.

The shadow price of forests *should be* an index of the scarcity of forests. Now forests absorb carbon emissions which are generated in the production of value added, using fossil fuels. In fact, forests may be seen as joint social capital required in the production of output and value added. This social capital has 'excess' capacity if emissions of carbon are less than the carbon sequestered by forests per year. The carbon sequestered is called uptake. When uptake exceeds emissions there is excess capacity, and some forested areas can be cut down for alternative uses. However, when emissions exceed uptake, forests become 'scarce' and the index of scarcity must be greater than one.

From the Maximum Principle, it follows that:

$$\dot{\gamma}(t) \leq 0 \quad \forall t_0 \leq t < T \tag{4.10}$$

from which it follows that:

$$\dot{\lambda}(t) \geq 0, \quad \forall t_0 \leq t < T \tag{4.11}$$

If the objective function is strictly concave, then the costate variable is *monotonic, decreasing*. Consequently,

$$\dot{\lambda}(t) > 0 \quad \forall t_0 \leq t < T \tag{4.12}$$

Hence $\lambda(t)$ is the correct index of scarcity, for it increases when the stock of forests decreases. If the objective function is strictly concave, then the index of scarcity is *monotonic, increasing.* This is of course a special case, as argued by Farzin (1992).

In order to obtain a cardinal index of scarcity, we make assumption 5. According to this assumption, the social value of value added (or output) is 'valued at par' only when there is no externality, that is when carbon emissions are just equal to the carbon sequestration by forests. Furthermore, if the value of value added declines proportionally with the excess of carbon emissions over carbon sequestration, then the *magnitude* of the externality is taken into account. This suggests the normalization of the shadow price to 1 when emissions equal uptake.

It follows that the shadow price is of the form:

$$\lambda(t) = \lambda\left(\frac{e(t)}{Q(t)}\right), \quad \text{where } \lambda(1) = 1 \tag{4.13}$$

where $e(t)$ are emissions of carbon in MT and $Q(t)$ is uptake of carbon in MT.

The above form of the equation yields a cardinal index of relative scarcity which over time need not necessarily be monotonic; monotonicity will hold only as a special case. For example, monotonicity is obtained if $e(t)$ is growing but $Q(t)$ is held constant, as in Model 1 in the next section.

Our econometric estimates in the next section use equation (4.13) to compute and estimate a cardinal measure of relativity scarcity over time. The interpretation is straightforward: when value added generates emissions greater than uptake, $\lambda(t) > 1$, that is forests are valued at a premium, and when emissions are less than uptake, $\lambda(t) < 1$, forests are valued at a discount, for in the latter case we can produce more emissions by producing more value added without adding carbon emissions into the 'Canadian' atmosphere.

As stated above, in general $\lambda(t)$ will be non-monotonic, but monotonicity can be obtained as a special case. It will be shown that this special case occurs when the uptake $Q(t)$ is held constant, and when the analysis is confined to trend analysis only. Such a *ceteris paribus* thought experiment is instructive as a first approximation in the valuation of forests (Model 1) below. When $Q(t)$ is not held constant, we get a different (non-monotonic) time path of the shadow price $\lambda(t)$. In the next section, two such models are considered, depending on the method of estimation, where Model 2 is a straight linear regression over time and Model 3 is an ARIMA model.

This completes the presentation of the model. The next section gives an econometric estimation of the main equation, which is equation (4.7), reproduced below:

$$\Delta x(t) = \lambda(t)\Delta W(t) \tag{4.7}$$

ECONOMETRIC ESTIMATES

Before presenting the estimations, it might be useful to consider Figure 4.1, which is a graph of Canadian carbon emissions in MT, as well as the sequestration of carbon by Canadian forests, for the period 1961–91. A time series of atmospheric carbon emissions for Canada over the period 1961–91 is provided by Marland *et al.* (1994) and published by CDIAC. These authors report a continuing growth in emissions from 1961 until 1980. There was a drop in emissions after 1980, and the pattern from 1980 to 1987 was erratic but with a small downward trend, as shown below. During the last two years of the series, emissions have decreased 7 per cent after peaking in 1989 at a level of 120.7 MTC. It is clear that Canadian emissions of carbon have been increasing, although the rate of growth has become volatile, with a slight downward trend from 1980 onwards. This slowdown is in part a legacy of the oil shocks of 1973 and 1979, and in part due to structural changes within the economy, with the relative decline of manufacturing in most industrialized nations. From about 1969, Canadian emissions began to exceed the amount of carbon sequestered by Canadian forests. (See Figure 4.1.)

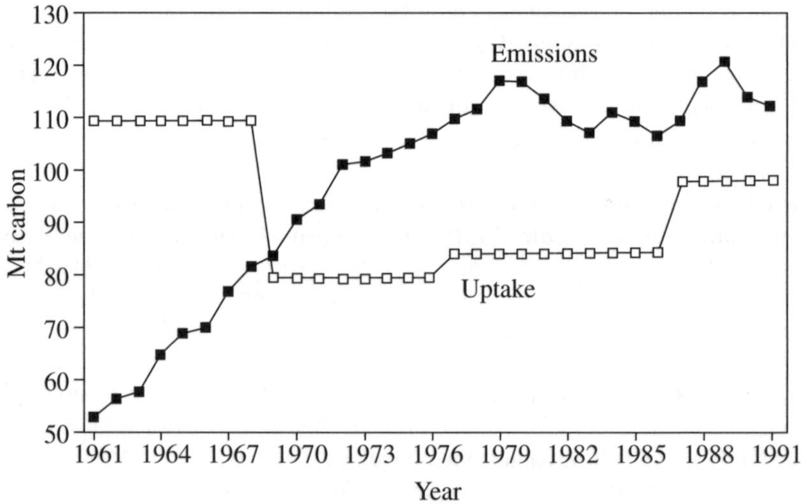

Figure 4.1 Canadian carbon emissions versus carbon uptake by Canadian forests

Data on carbon sequestration (or uptake) is only just beginning to be refined, a subject that is itself quite complex. Details on the data on carbon sequestration are considered in Appendix 1. We now proceed with the presentation of our empirical results. As stated above, we report three model estimates.

Model 1: Trend Analysis with Constant Uptake

Model 1 is a first approximation, obtained by fitting a time trend to emissions, holding uptake constant. A time trend can be fitted to Canadian manufacturing value added for the period 1961–91.

Emissions

A linear regression of carbon emissions on time is fitted to annual Canadian data. In light of the comments made by Marland *et al.* (1994), we are interested in testing for a structural break in 1980. The possibility of such a break is intuitively appealing, given the oil price shock that occurred in 1979 and the structural changes in manufacturing considered to have taken place following the 1980–82 recession. The test employed is based on a dummy variable technique.

The basic model is

$$\text{Emissions}_t = \alpha + \beta t + \mu_t$$

We define the dummy variable as follows:

$D_1 = 1$ for the period 1980 onward,
$D_1 = 0$ otherwise.

To test whether the structures for the two periods (prior to 1980, and the period after 1980) are different, the models assume the following specification:

$$\alpha = \alpha_0 + \alpha_1 D_1 \quad \beta = \beta_0 + \beta_1 D_1$$

Thus the unrestricted model is:

$$\text{Emissions}_t = \alpha_0 + \alpha_1 D_1 + \beta t + \beta_1 (D_1 t)$$

The estimated fitted models are,

prior to 1980: $$\text{Emis}\hat{\text{s}}\text{ions}_t = \hat{\alpha} + \hat{\beta}_0 t$$

1980 onward: $$\text{Emis}\hat{\text{s}}\text{ions}_t = \hat{\alpha}_0 + \hat{\alpha}_1 + (\hat{\beta}_0 + \hat{\beta}_1)t$$

Thus a failure to reject the null hypothesis that $\alpha_1 = \beta_1 = 0$ would indicate the presence of a structural break at 1980.

A Lagrange multiplier test on the residuals following a first run of the model indicated the presence of first-order serially correlated errors (p-value = 0.0080). Based on White's test for heteroscedasticity, we reject the null hypothesis that the errors have non-constant variance (p-value = 0.1218). The regression was estimated again, with adjustments to the standard errors of the estimated coefficients (based on the estimation of a consistent covariance matrix).

The fitted model is,

prior to 1980:

$$\hat{\text{Emissions}}_t = 50.76 + 3.63t$$
$$(0.17)\ (0.43)$$
$$t = (34.35)\ (21.91)$$
$$R^2 = 0.998$$
$$\text{d.f.} = 27$$

1980 onward:

$$\hat{\text{Emissions}}_t = 54.60 + 55.56 + (3.63 - 3.39)t$$
$$(11.11) \qquad (0.44)$$
$$t = \quad (4.32) \qquad (-4.69)$$
$$\text{RMSE} = 3.60$$

That is,

$$\hat{\text{Emissions}}_t = 110.16 + 0.24t$$

A joint significance test of the null hypothesis that $\alpha_1 = \beta_1 = 0$ leads to rejection of the null (p-value = 0), and we conclude that there was a structural break in 1980.

Value added

Statistics Canada provides us with a time series of value added in Canadian manufacturing. Value added has essentially continued to grow over time; however, beginning in about 1980, levels of value added have become more variable. Of course, the period 1980–91 includes two major disturbances, the deep recession of 1981–82 and the onset of the lesser recession that began in 1990.

A linear time trend is fitted to annual data on value added in Canadian manufacturing. The model is simply:

$$\text{Value Added}_t = \alpha + \beta t + \mu_t$$

As in the case of carbon emissions, the results of a first run of the model indicated the presence of first-order serially correlated errors (p-value = 0.0088). White's test indicates the presence of heteroscedastic errors (p-value = 0.0060). The model was estimated again, with adjustments made to the standard errors as above.

The fitted model is:

$$\text{Added}\hat{\ }\text{Value}_t = 46.90 + 2.01t$$
$$(1.97)\ (0.17)$$
$$t = (23.74)(11.48)$$
$$R^2 = 0.9954$$
$$\text{RMSE} = 5.67 \qquad \text{d.f.} = 29$$

The value of the forests

The estimated shadow price is shown in Figure 4.2.[3] Note that, except for the structural break, a monotonic shadow price is obtained as (a) uptake is held constant at 89 MTC, and (b) a linear trend is fitted. While this is a special case, it is nevertheless instructive mainly as a thought experiment. The estimated value of Canadian forests is then given in Figure 4.3, with the dotted lines giving the 95 per cent confidence bands based on the calculated confidence interval for the individual in-sample point forecasts, or individual predictors. For a comparative perspective, consider Figure 4.4, where the estimated value of the forests is expressed as a percentage of GDP. Note that the 1990 figure of the value of the forests as a proportion of real GDP is close to 23 per cent. But the important estimate is the value of forests per hectare: this is the *marginal social opportunity cost of one hectare of forest.* The value of the forests per hectare is shown in Figure 4.5. The 95 per cent confidence intervals yield the 'high' and the 'low' estimates.

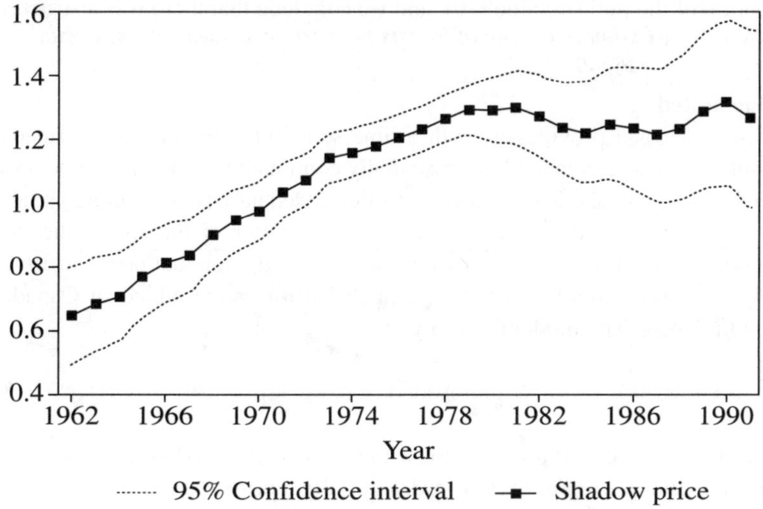

Figure 4.2 First approximation of the shadow price

Model 2: Trend Analysis with Variable Uptake

Next we relax the assumption of a constant uptake. Here we also exploit the information in the residuals and estimate autocorrelation models, using Cochrane–Orcutt as the method for estimation.

The fitted model is,

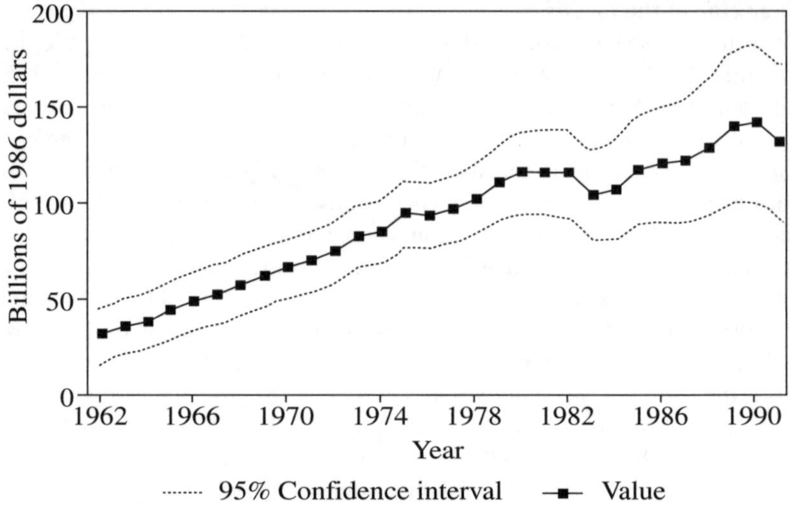

Figure 4.3 *Estimated value of forests for carbon sequestration, Canada, 1962–91*

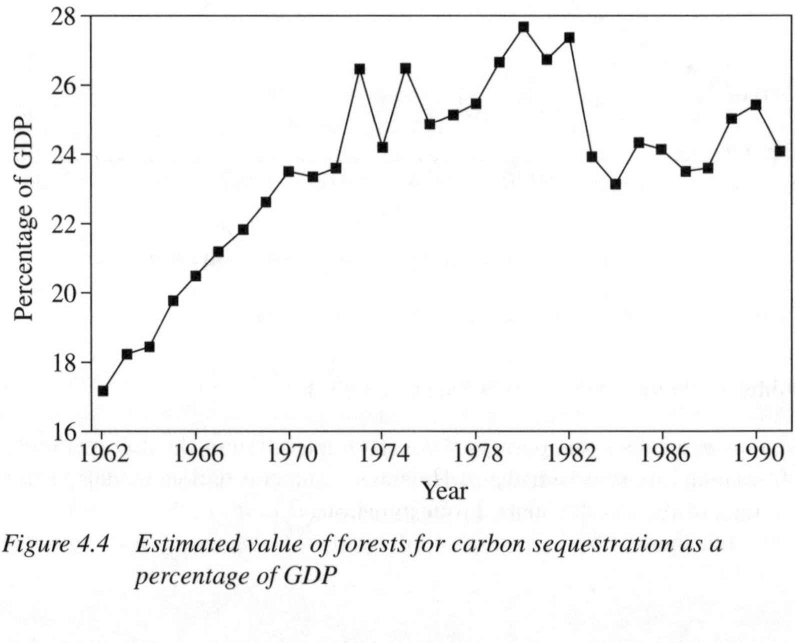

Figure 4.4 *Estimated value of forests for carbon sequestration as a percentage of GDP*

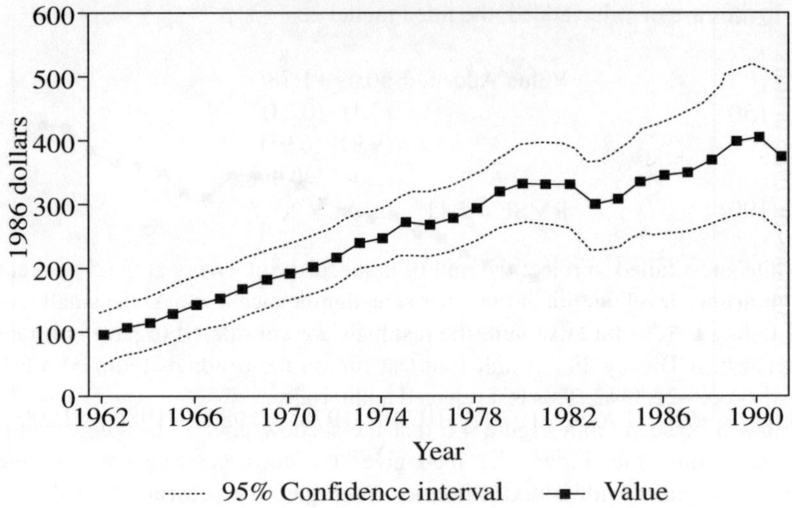

Figure 4.5 Forest value per hectare, Canada, 1962–91

prior to 1980:

$$\begin{aligned} \hat{\text{Emissions}}_t &= 54.6 + 3.26t \\ &\quad (5.94)\ (0.43) \\ t &= (9.19)(7.63) \\ R^2 &= 0.9992 \\ \text{d.f.} &= 25 \end{aligned}$$

1980 onward:

$$\begin{aligned} \hat{\text{Emissions}}_t &= 54.60 + 66.38 + (3.26 - 3.56)t \\ &\qquad\quad (15.36) \qquad\qquad (0.76) \\ t &= \qquad (4.32) \qquad\qquad (-4.69) \\ \text{RMSE} &= 3.05 \end{aligned}$$

That is,

$$\hat{\text{Emissions}}_t = 120.98 - 0.30t$$

A Dickey–Fuller unit root test on the residuals returned a test statistic value of –24.6159, leading to the rejection of a unit root. A joint significance test of the null hypothesis that $\alpha_1 = \beta_1 = 0$ leads to rejection of the null (p-value = 0.0003), and we conclude that there was a structural break in 1980. In fact, contrary to the interpretation given by Marland *et al.* (1994), these results show that there is a slight downward trend in Canadian carbon emissions since 1980.

In the case of value added, the fitted model is:

$$\begin{aligned} \text{Value\^{}Added}_t &= 50.59 + 1.78t \\ &\quad\ (5.1) \quad (0.21) \\ t &= (9.91)\ (6.91) \\ &\qquad R^2 = 0.9965 \\ \text{RMSE} = 5.11 &\qquad \text{d.f.} = 27 \end{aligned}$$

White's test failed to reject the null of homoscedastic errors at the 5 per cent significance level but not at the 1 per cent significance level. At the small risk of failing to reject a false null, the residuals are considered to have constant variance. A Dickey–Fuller unit root test run on the residuals returned a test statistic of –22.3342. This test rejected a unit root.

It will be seen from Figure 4.6 that the shadow price is no longer completely monotonic. Figure 4.7 then gives the corresponding value of the forests, together with confidence intervals. Figure 4.8 presents the value of the forests as a proportion of real GDP and Figure 4.9 gives the value, in constant 1986 dollars, of a hectare of forest. For instance, the high and low estimates for 1986 and 1990 are given in Table 4.1.

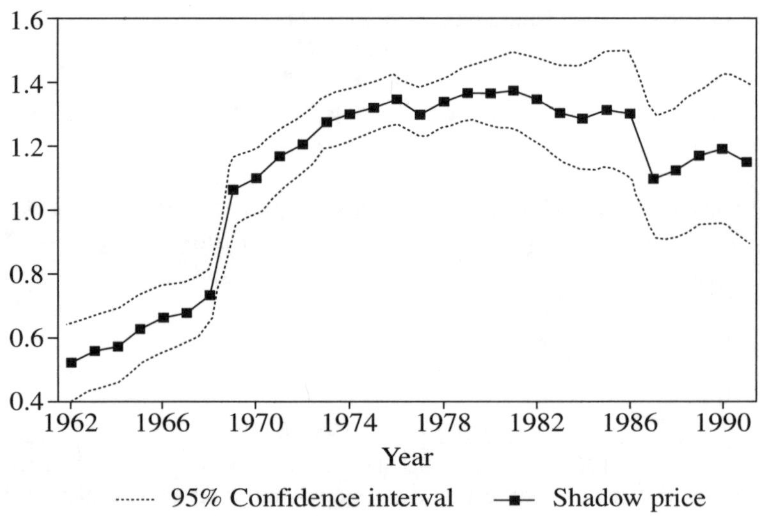

Figure 4.6 An ARIMA estimate of the shadow price

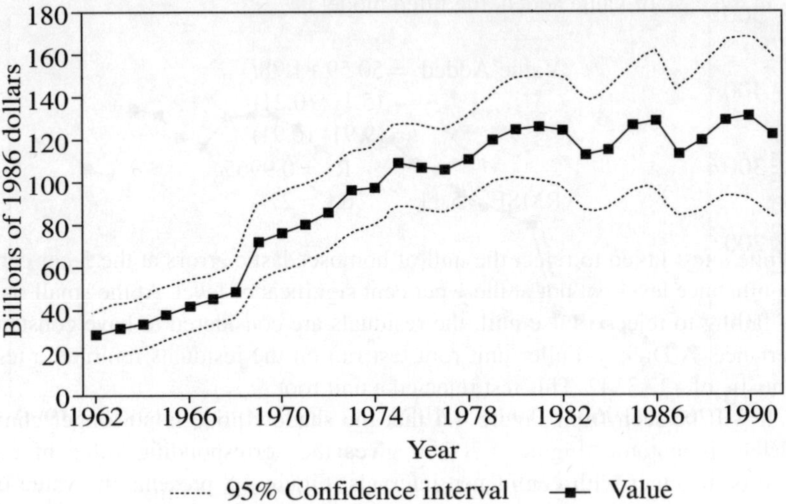

Figure 4.7 An ARIMA based estimate of the value of forests for carbon sequestration, Canada, 1962–91

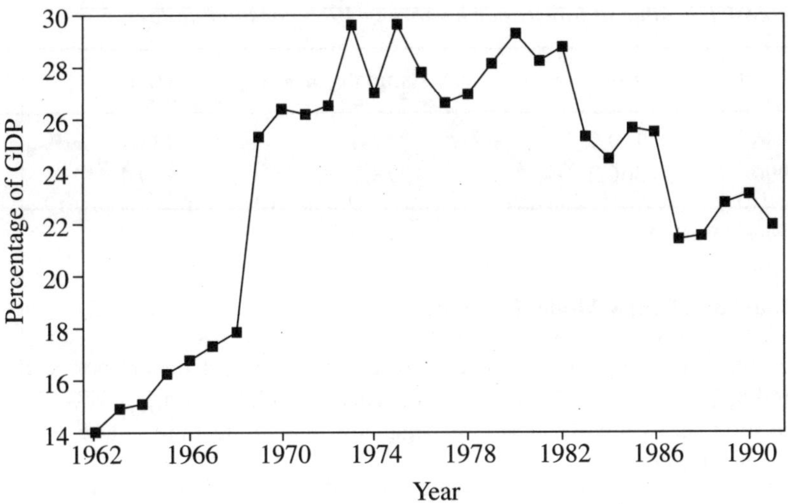

Figure 4.8 An ARIMA based estimate of the value of forests for carbon sequestration as a percentage of GDP

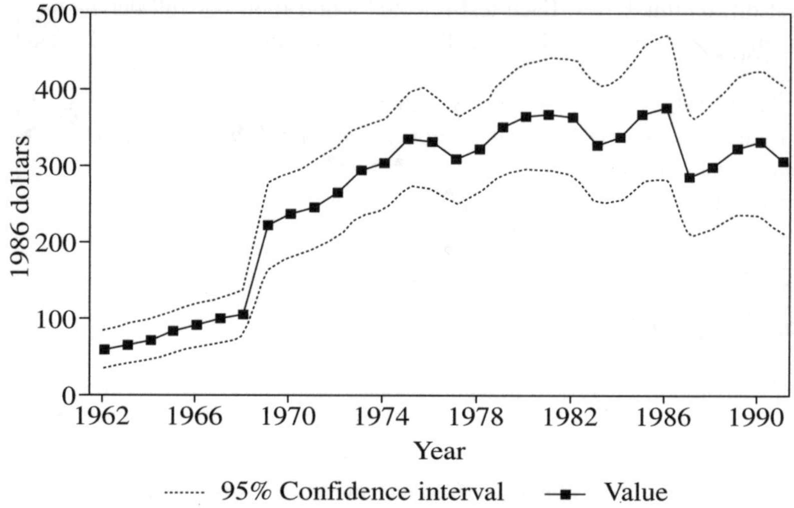

Figure 4.9 *An ARIMA based estimate of forest value per hectare, Canada, 1962–91*

Table 4.1 *Value of forests per hectare in 1986 constant dollars*

	Low	Estimate	High
1986	259.9	347.0	434.1
1990	266.2	378.3	490.4

Source: Figure 4.9.

Model 3: ARIMA Model Estimates

An alternative approach involves fitting a model based on the notion that present levels of emissions and value added are related to past levels. Since the time series are so short, we began by first fitting pure autoregressive models of order *n*.

Carbon emissions

In the case of carbon emissions, a first-order autoregressive model was fitted to the data. An LM (Lagrange multiplier) test on the residuals indicated the presence of first order serially correlated errors (p-value = 0.0014). Consequently, the series was first differenced to make it stationary. Three alternative unit root tests were carried out (Dickey–Fuller, Phillips–Perron, and Stock–

Watson) on this series. Each test rejected a unit root. For full details on the justification of the ARIMA modelling in this chapter, see details of the unit root tests given in Appendix 2.

An ARIMA (1, 1, 0) model was then fitted to annual data on Canadian carbon emissions. (Higher-order autoregressive terms were found to be statistically insignificant.)

The fitted model is:

$$E'_t = 0.4606E'_{t-1}$$

where $E'_t = (1 - B)E_t$

or $(1 - 0.4606B)(1–B)\ E_t = a_t$ where a_t is the white noise error term
(0.1660)
t = (2.7744)
d.f. = 28
RMSE = 3.7192

Value added

In the case of value added, a first-order autoregressive model was also fitted to the data. An LM test on the residuals indicated the presence of first-order serially correlated errors (p-value = 0.0304) and, therefore, this series too was first differenced to make it stationary. Three alternative unit root tests were carried out (Dickey–Fuller, Phillips–Perron and Stock–Watson) on this series. Each test rejected a unit root at the 5 per cent level. (See Appendix 2 for details.) The model based on first differences was then estimated. Again additional autoregressive terms were found to be statistically insignificant; however, the coefficient on the dependent variable was found to be statistically insignificant. For this reason the series was second differenced, with unit root tests rejecting the presence of a unit root as above. An ARIMA (1, 2, 0) model was then fitted to annual data on value added in Canadian manufacturing.

The fitted model is:

$$VA'_t = -0.4254VA'_{t-1}$$

where $VA'_t = (1 - B)^2\ VA_t$

or $(1 + 0.4254\text{B})\ (1–B)^2\ VA_t = a_t$, where a_t is the white noise error term.
(0.1756)
t = (–2.4224)
d.f. = 27
RMSE = 6.3613

Value of the forests

Figure 4.10 gives the estimated shadow price with the 95 per cent confidence intervals. Figure 4.11 gives value of the forests, again with the confidence

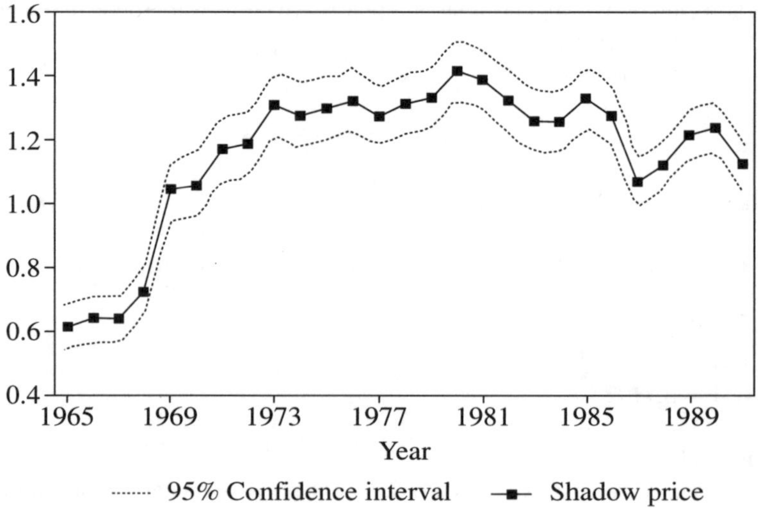

Figure 4.10 Final ARIMA based estimated shadow price

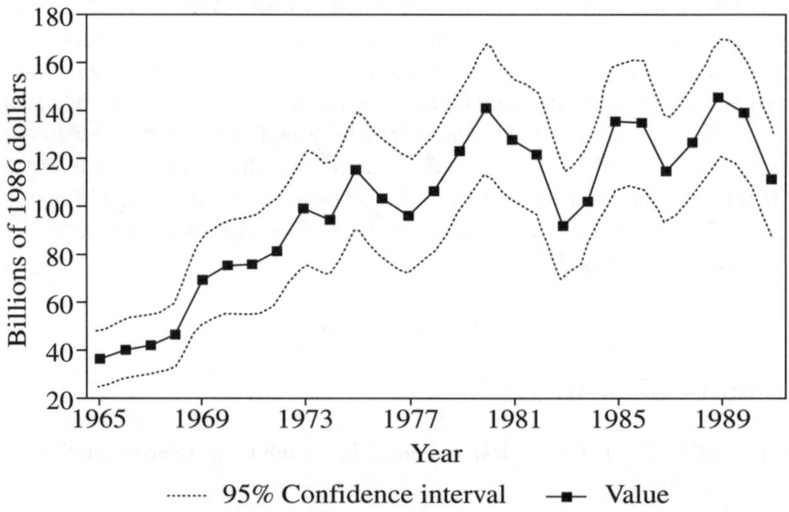

Figure 4.11 Final ARIMA based estimated value of forests for carbon sequestration, Canada, 1965–91

intervals. Figure 4.12 gives the value of the forests as a proportion of GDP, and Figure 4.13 gives the value of the forests per hectare in 1986 constant dollars. The high and low estimates of 1986 and 1990 are given in Table

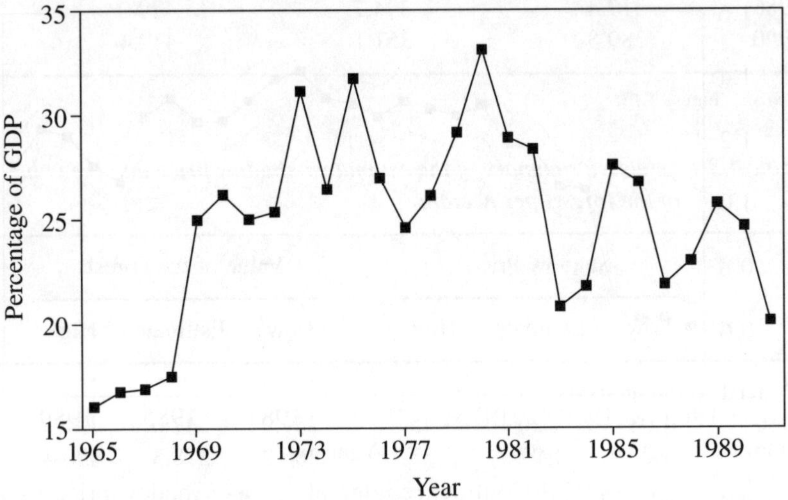

Figure 4.12 Final ARIMA based estimated value of forests for carbon sequestration as a percentage of GDP

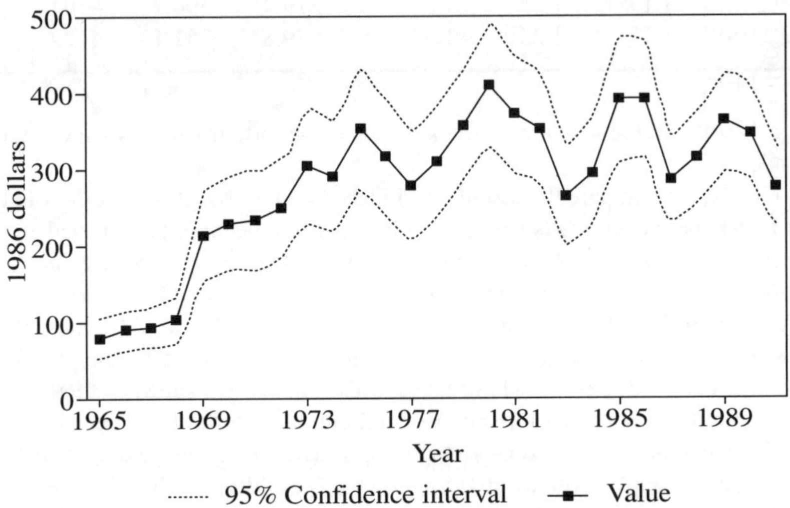

Figure 4.13 Forest value per hectare, Canada, 1965–91

Table 4.2 Value of forests per hectare in 1986 constant dollars

	Low	Estimate	High
1986	319.4	394.7	470.0
1990	289.8	351.1	412.4

Source: Figure 4.13.

Table 4.3 Selective summary of the estimates: shadow price and the value of the forests per hectare

	Shadow Price			Value of the Forests		
	Low	Estimate	High	Low	Estimate	High
Model 1: trend analysis						
1986	1.08	1.26	1.45	259.9	347.0	434.1
1990	1.02	1.27	1.53	266.2	378.3	490.4
Model 2: trend analysis						
1986	1.10	1.31	1.51	280.5	376.5	472.5
1990	0.97	1.20	1.43	233.6	324.2	414.7
Model 3: ARIMA model						
1986	1.19	1.29	1.38	319.4	394.7	470.0
1990	1.17	1.25	1.33	289.8	351.1	412.4

4.2. Finally, Table 4.3 summarizes the most important results of the three models.

Finally, we compare the results of Models 2 and 3. Note first that the value of forests per hectare falls between 1986 and 1990, because the forested area goes up. The 1990 ARIMA estimate (Model 3) is higher because the in-sample forecast for emissions is higher, which is in turn higher due to a higher emissions level in the previous year (1989). We would expect the ARIMA model to pick this up. However, from Figure 4.14 it is clear that the two models yield fairly close estimates. This is the case even when there was a structural break incorporated in Model 2 but none in Model 3, because the ARIMA estimation procedure reflects the structural changes endogenously: the fitted in-sample estimate of 1981 shows a sharp 'drop'.[4] The fall shows up in 1981 because the ARIMA model is of course an (autoregressive) distributed lag model.

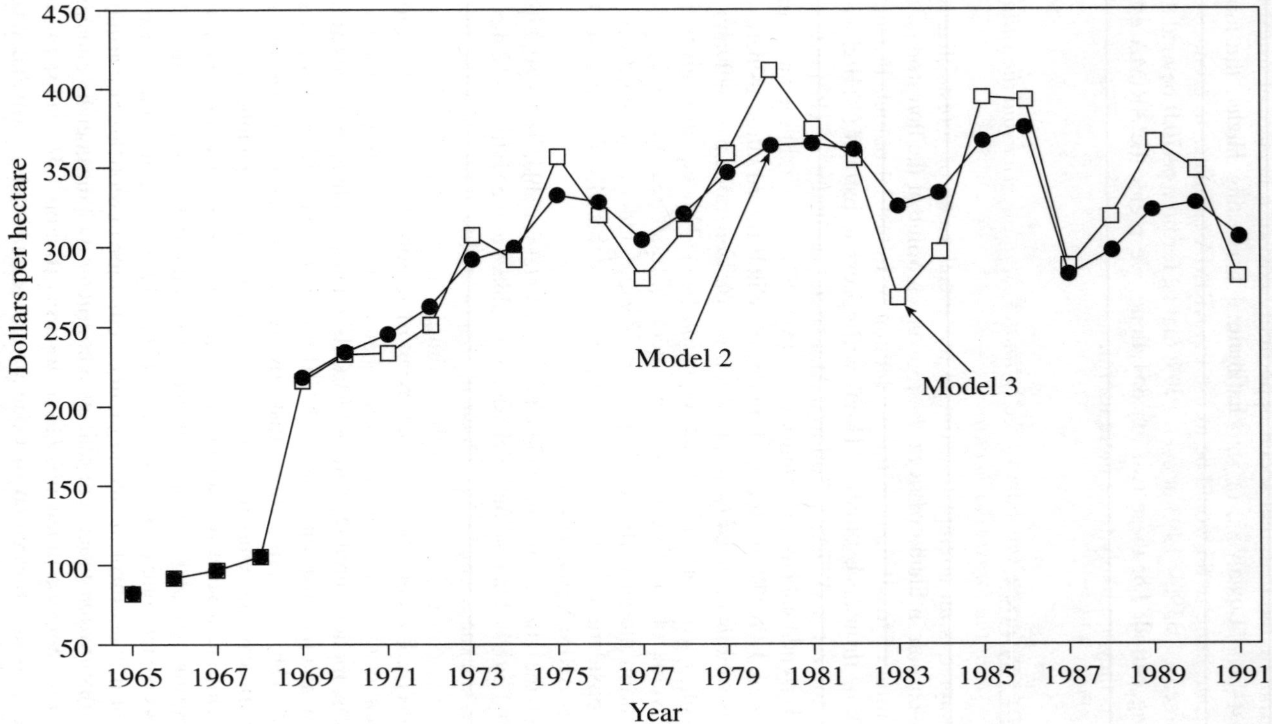

Figure 4.14 Value of Canadian forests for carbon sequestration 1965–91

The ARIMA model may indeed be a better forecasting tool, for if we were to incorporate the marginal social opportunity cost of a hectare of forest in any cost–benefit analysis, or into an optimal harvesting model, the most appropriate thing to do would be to use the ARIMA estimates to produce ex ante forecasts, say for 1993 or even 1995; that is, for time periods beyond the sampling period. For these two reasons alone, we prefer the ARIMA estimates, as they are likely to be more useful.

CONCLUSIONS

This chapter is an attempt to estimate the social value of forests that sequester carbon in the biomass of the forests, carbon that is a by-product of the burning of fossil fuels in the production of industrial output. It avoids the lack of invariance associated with willingness to pay (WTP) and willingness to accept (WTA) as follows. At a social optimum, WTP must also reflect marginal social opportunity costs (MSOC), so that, at the optimum, MSOC equals WTP. We set up a model in which the equation to be estimated as well as the shadow price is in principle generated from a dynamic optimizing framework. This is our equation (4.7). This equation gives the MSOC value of the forests. As the objective functional in the dynamic optimization model is the maximization of value added, with no distributional considerations, the implied numeraire in the MSOC value of the forests is indeed value added.

While the shadow price, which is the costate variable, is in principle generated by the dynamic optimization, it is subject to the restrictive complementary slackness condition, which in theory yields a zero shadow price whenever a constraint in not fully binding. The result is that the costate variable is not a complete cardinal index of scarcity. As we are using the reciprocal of the costate variable as the shadow price of the forests, the same limitation applies to the shadow price of the forests. To overcome this problem, we assume that value-added is valued at par only when there is no externality, that is when carbon emissions are equal to carbon sequestered by the forests. We thus make the shadow price a function of the ratio of carbon emissions to sequestration. When emissions are just equal to sequestration, the value of the forests is just equal to the value added made possible by forests as social (overhead) capital. Consequently, at that time value added is valued at par. However, as soon as carbon emissions exceed carbon sequestration, the marginal social value of value added is less than its constant dollar value, because of the externality. Reciprocally, the value of the forests is higher than the constant dollar value of value added. In this formulation the shadow price is a cardinal index of the relative scarcity of the forests, which are really a joint input in the production process that uses carbon-emitting fossil fuels.

The econometric estimates of the shadow price of the forests and the resulting MSOC value of forests per hectare are obtained by a time trend model (Model 2) and an ARIMA model (Model 3). We regard the ARIMA model estimates to be more representative of current conditions as the time trend model (Model 2) merely extrapolates from an established trend, even when structural breaks are taken into account. The estimated shadow price of the forests (in 1990) is a premium of the order of 20 per cent. The corresponding MSOC value of a hectare of forest is between \$290 and \$412 in 1986 constant dollars. Naturally, the estimates are subject to the limitations of the data.

ACKNOWLEDGMENTS

This chapter was first presented as a paper at the Annual Conference of Canadian Resource and Environmental Economics, Simon Fraser University, Vancouver, BC, 30 September–1 October 1995. We are grateful to Dr Bill Veloce (Brock University) and Dr Patricia Roberts-Pichette (Environment Canada) for helpful comments on an earlier version, but the usual disclaimer applies. This research is funded by the Social Sciences and Humanities Research Council of Canada, grant # 410-94-1121.

NOTES

1. Actually, along the optimal solution,

$$\frac{\partial w^*}{\partial x^*} = \gamma^*(t)$$

   but we are only interested in the first component of the vector $\boldsymbol{x}$. The equation given above is a standard property of the maximum principle. Chiang (1992, pp. 206–7) states it without proof, but Leonard (1987) gives an explicit proof of this property.
2. In the absence of indirect taxes, the sum of value added is nothing but GDP.
3. In all the figures in this chapter, the dashed lines give the 95 per cent confidence intervals.
4. Strictly speaking, a structural break could be introduced if the sample were split into two: a pre-1980 sample and a post-1980 sample. But this would give a sample of 10 observations, requiring the estimation of four parameters. This would drastically reduce the degrees of freedom. In any case, as argued in the text, we do not believe that a structural shift is in any way biasing our ARIMA results. To test the robustness of our results, we estimated an alternative model (Model 4) which was ARIMA (0,1,1) which showed no structural break and was statistically significant. However, the numerical results were virtually identical to those of Model 3. For this reason, the results of Model 4 are not reported in this chapter, but they do confirm the robustness of our numerical estimates

REFERENCES

Barnett, H.J. (1979), 'Scarcity and Growth Revisited', in V.K. Smith *Scarcity and Growth Reconsidered*, Baltimore: Johns Hopkins University Press, pp. 163–217.

Barnett, H.J. and C. Morse (1963), *Scarcity and Growth: The Economics of Natural Resource Availability*, Baltimore: Johns Hopkins University Press (for Resources for the Future).

Brody, A. (1970), *Proportions, Prices and Planning*, Amsterdam: North Holland.

Brown, G.M. and B.C. Field (1978), 'Implications of alternative measures of natural resource scarcity', *Journal of Political Economy*, 86 (2), pt.1 (April), 229–43.

Chiang, A.C. (1992), *Elements of Dynamic Optimization*, New York: McGraw-Hill.

Dore, M.H.I. (1995), 'Dynamic Games in Macro Models: a Critical Appraisal', *Journal of Post-Keynesian Economics*, 18 (1), Fall, 107–23.

Dore M.H.I. and A.J. Ward (1995), 'Modelling Sustainable Development: a Nonlinear Analysis', Brock University, mimeo.

FAO (1995), 'Forest resources assessment 1990; Global Synthesis', FAO Forestry Paper, **124**.

Farzin, Y.H. (1984), 'The effect of discount rate on depletion of exhaustible resources', *Economic Journal*, 92 (October), 841–51.

Farzin, Y.H. (1992), 'The time path of scarcity rent in the theory of exhaustible resources', *Economic Journal*, 102 (July), 813–30.

Fisher, A.C. (1979), 'Measures of Natural Resource Scarcity', in V.K. Smith, *Scarcity and Growth Reconsidered*, Baltimore: Johns Hopkins University Press, pp. 249–75.

Hall, D.C. and J.V. Hall (1984), 'Concepts and measures of natural resource scarcity with a summary of recent trends', *Journal of Environmental Economics and Management*, 11, 363–79.

Hanson, D.A. (1980), 'Increasing extraction costs and resources prices: some further results', *Bell Journal of Economics*, 11 (Spring), 335–42.

Heal, G.M. (1976), 'The relationship between price and extraction cost for a resource with a backstop technology', *Bell Journal of Economics*, 7 (Autumn), 371–8.

Horbulyk, T.M. and W.A. Ross (1993), 'Potential to Sequester Carbon in Canadian Forests: a Comment', *Canadian Public Policy*, XIX(4), 463–8.

Johnston, M.H., F.W. Bell and J.T. Bennett (1980), 'Natural resource scarcity: empirical evidence and public policy', *Journal of Environmental Economics and Management*, 7 (September), 256–71.

Kurz, W.A., M.J. Apps, T.M. Webb and P.J. McNamee (1992), 'The carbon budget of the Canadian forest sector: Phase1', Forestry Canada, Information Report NOR-X-326 (Edmonton: Forestry Canada, Northwest Region).

Leonard, D. (1987), 'Co-state variables correctly value stocks at each instant: a proof', *Journal of Economic Dynamics and Control*, 11, 117–22.

Marland, G., R.J. Andres and T.A. Boden (1994), 'Global, regional and national CO_2 emissions', in T.A. Boden, D.P. Kaiser, R.J. Sepanski and F.W. Stoss (eds), *Trends '93: A Compendium of Data on Global Climate Change*, ORNL/CDIAC-65, Carbon Dioxide Information Analysis Center, Oak Ridge National Laboratory, Oak Ridge Tennessee.

Nordhaus, W.D. (1974), 'Resources as a constraint on growth', *American Economic Review*, 64, (May), 22–6.

Ramanathan, R. (1992), 'Introductory econometrics with applications', London: Dryden, with Harcourt Brace Jovanovich.

Slade, M.E. (1982), 'Trends in natural-resource commodity prices: an analysis of the time domain', *Journal of Environmental Economics and Management*, 9 (June), 122–37.
Smith, V.K. (1979), *Scarcity and Growth Reconsidered*, Baltimore: Johns Hopkins University Press (for Resources for the Future).
Solow, R.M. and F.Y. Wan (1976), 'Extraction costs in the theory of exhaustible resources', *Bell Journal of Economics*, 7 (Autumn), 359–70.
van Kooten, G.C., Louise M. Arthur and W.R. Wilson (1992), 'Potential to Sequester Carbon in Canadian Forests: Some Economic Considerations', *Canadian Public Policy,* XVIII(2), 127–38.
van Kooten, G.C., W.A. Thompson and I. Vertinsky (1993), 'Economics of Reforestation in British Columbia when Benefits of CO_2 Reduction are taken into Account', in W. Adamowicz, W. White and W.E. Phillips (eds), *Forestry and the Environment: Economic Perspectives*, Wallingford, UK: Commonwealth Abstract Bureau International.

APPENDIX 1: DETERMINATION OF THE AMOUNT OF CARBON SEQUESTERED AT TIME *T*

Essentially, the net flux of carbon (*C*) between the forest ecosystem and the atmosphere determines whether forests are part of the problem or part of the solution with regard to changes in the concentration of CO_2 in the atmosphere. We are interested in obtaining the most appropriate estimate of carbon uptake by forest biomass reported in the scientific literature. In their examination of the potential role of reforestation policies in offsetting Canada's emissions of atmospheric carbon dioxide, van Kooten *et al.* (1992, pp. 127–38) determine the rate at which carbon is sequestered on Canada's forested lands by multiplying forest land area (ha) in each ecoclimatic province by the annual productivity of the forest in each zone (m^3/ha/yr) and by the average amount of carbon in green timber (260 kg/m^3). The results of this calculation are 203 million tonnes (MT) per year of carbon sequestered by Canadian forests. However, there are problems with using an average carbon content value, in that it is too general. Consequently, van Kooten *et al.* (1993) employ a different set of assumptions to estimate a lower rate of carbon sequestration on the same forest lands. The new value is 157.1 MT per year, representing a decrease of 23 per cent relative to that arrived at in van Kooten *et al.* (1992). This decrease is due to the use of specific carbon content of wood estimates for the mix of tree species found in each ecoclimatic province. This change in approach effectively reduces the average rate of carbon sequestration from 260 kg/m^3 to 200 kg/m^3. However, this figure (157.1 MT per year) is probably an overestimate. Although a key criticism made by Horbulyk and Ross (1993) of the figure obtained by van Kooten *et al.* (1993) has been taken into account in the revised estimate, Horbulyk and Ross also criticize van Kooten *et al.* for their use of mean annual increment data for average forest productivity as if they were net values from which deductions had been made for all mortality losses due to fire and pest epidemics. Horbulyk and Ross make the important point that, where losses occur, forests are not only unable to sequester carbon at average rates but also release atmospheric carbon. New estimates for Canada of the 1990 carbon uptake by forest biomass and forest soils are revealed in Kurz *et al.* (1992). Given the scope of the analysis, we believe that their estimates are the best available to date. These data are used to approximate our (4.8) and (4.9).

The results of the standard simulation run show that, prior to accounting for disturbances, the biomass C pools of Canadian forest ecosystems sequestered an estimated 109.3 MT of carbon, comprising the 92 MT increase in the biomass C pool and the 17.3 MTC transferred to the soil C pools as detrital material (litter fall and tree mortality). Carbon uptake is distributed across all forests in Canada (Kurz *et al.*, pp. 41–4).

Every year a small proportion of the Canada forests is disturbed by fires, insects or harvesting. In the reference year 1986, disturbances released 20.3 MT of the carbon sequestered in forest biomass directly to the atmosphere. Disturbances also transferred biomass carbon to the forest products sector (44.2 MTC harvested biomass) and to the soils pools (55.4 MTC). The latter transfer consisted of slash and dead standing trees after harvesting, fires and insect-induced tree mortality (included in the soil C pools in this model).

Thus we conclude that the biomass carbon pools of the Canadian forest ecosystem sequestered approximately 109.3 – 20.3 = 89 MT of carbon. Carbon released through disturbance but transferred to the forest products sector or soils pools is not deducted, as it effectively remains sequestered. Kurz *et al.* caution that this is a single time step simulation and that extrapolation would be inappropriate. However, to facilitate the analysis of carbon value, several assumptions are made that allow us to create a time series of carbon uptake by Canadian forests, however imperfect. We assume that species composition and age distribution are fairly constant over the period 1961–91. We denote uptake at time t as $Q_t = q \cdot a_t$, where q = rate of carbon sequestration = 89 MTC, as calculated above, and a_t = forest area at time t. Kurz *et al.* identify 1986 as the reference year for the model simulation, utilizing data from the period 1981–9. For example, estimates of carbon released due to wildfire are based on the 10-year average of area burned. Data on the extent of the forest (a_t) is sketchy since forest inventories are done only intermittently. The period 1961–91 saw the extent of the forests drop from about 443 million to 322 million hectares through the 1970s and return to about 400 million hectares towards the end of the 1980s. Based on these data, the 9-year average (1981–9) of forest area is about 361 million hectares. Thus we arrive at a value for q of 89/361 = 0.2465 MTC or about one-quarter of a megatonne for every million hectares of forest. This shows how we obtained a reasonable approximation for our equations (4.8) and (4.9), given the limited availability of the data.

APPENDIX 2: UNIT ROOT TESTS

Emissions

In general, econometric methods, including ARIMA modelling, are valid only if the time series being modelled is stationary. We took first differences of emissions based on the fact that the original time series of emissions was non-stationary. We determined this through a casual inspection of the correlogram, à la Box–Jenkins. However, we can also use the augmented Dicky–Fuller (ADF) tests: the T(rho-1) test and the regression *t* test. The T(rho-1) test statistic on the original emissions series turns out to be –3.72, while the critical value at 5 per cent is –21.7. As –3.72 > –21.7, we accept the null that there is a unit root. In terms of the regression *t* test, the statistic returned is –1.31, and the critical value is –3.41. Thus we again accept the null that there is a unit root. Therefore the results indicate that the series must be differenced. The next question is the determination of the number of autoregressive terms that will be necessary. We tried including additional lagged terms, but this proved to be not significant. We conclude that one lagged term is all that is required.

The T(rho-1) test statistic returned on the (first) differenced series is –24.32, whereas the critical value at 5 per cent is –21.7. Therefore we reject the null and conclude that there is no unit root in the first differenced series.

Similarly, in terms of the *t* test, the test statistic is –4.25 but the critical value is –3.41 at 5 per cent. Again we reject the null. Thus we have shown that the first differenced series is stationary and we do not have any of the problems of an explosive error term as described by Ramanathan (1992).

Value Added Series

When ADF is run on the original series, the T(rho-1) test statistic is –15.86, whereas the critical value is –21.7 at 5 per cent. In terms of the *t* test, the *t* value is –3.03, whereas the critical value is –3.41. Thus we accept the null of a unit root and conclude that the series must be differenced. Again the question is the determination of the number of autoregressive terms that will be necessary. We tried including additional lagged terms, but this proved to be not significant. We conclude that one lagged term is all that is required.

The T(rh0-1) test statistic returned on the (first) differenced series is –23.66, whereas the critical value is –21.7 at 5 per cent. We would conclude that there is no unit root. However, in the case of the t test, the test statistic is –2.72, whereas the critical value is –3.41. Based on this test, we would expect that there *is* a unit root. We therefore took second differences. When ADF is run on the second differenced series, the t test statistic returned is –4.62 (whereas the critical value is –3.41). We conclude that second differencing is required to yield a stationary series. As in the case above, we do not have the problem described by Ramanathan of an explosive error term. However, in this case, second differencing was required because first differencing did not yield a significant t statistic on the independent variable, that is value added in the last period. This appendix justifies the differencing of the two series, based on unit root tests.

5. The carbon cycle and the value of forests as a carbon sink: a tropical case study

Octavio A. Ramirez[1]

INTRODUCTION

Traditionally, forests have been perceived as a source of timber, wood and other extractive products such as medicinal and ornamental plants. In the past, the environmental services provided by forests have been neglected. This perspective has changed considerably due to international and local initiatives that view forests as ecosystems that render both productive and service functions. The latter include soil conservation/regeneration, water flow regulation/storage/cycling, and recreational and biodiversity reserve services.

Forest areas also play an important role in the biogeochemical cycles at the biosphere level, in particular the global carbon cycle (Dixon *et al.*, 1994). This cycle affects the atmospheric concentration of carbon dioxide (CO_2), a key gas in the greenhouse effect. The growing concentration of CO_2 in the atmosphere contributes to global warming and, therefore, to climate change (Brown, 1996). Forests store large amounts of carbon in their vegetation and soil and constantly exchange it with the atmosphere through photosynthesis and respiration. Forests can be a net source of atmospheric carbon because of human intervention or natural causes such as fires, logging, and the slash and burn practice used to clear land for agricultural production and cattle ranching. Net absorption of atmospheric CO_2 takes place during the regrowth that occurs after an intervention. Trees and other plants capture CO_2 from the atmosphere through photosynthesis, fixing carbon in their biomass and releasing oxygen (O_2).

Forest ecosystems contain 20 to 100 times more carbon per unit area than agricultural ecosystems (Andrasko, 1990; Schroeder *et al.*, 1993). When humans intervene in the forests, they modify the magnitude of the carbon reserves and induce changes in the circulation of this element, altering the carbon cycle and possibly affecting the climate (Brown, 1996).

The carbon sinking capacity of forests has recently become an internationally recognized environmental service. The use of forests to mitigate the atmospheric CO_2 concentration was proposed by the Intergovernmental Panel on Climate Change (IPCC). The United Nations Convention on Climate Change, a result of the Rio Conference, has been signed by 162 countries and ratified by 20. Nearly all these countries have agreed to reduce their CO_2 to 1990 levels by the year 2000 (Andersen, 1996). To comply with these commitments, at a meeting in Berlin in 1995, they established a pilot 'joint implementation' programme.

Joint implementation (JI) refers to cooperative initiatives between two or more governments with the aim of sequestering CO_2 currently in the atmosphere or reducing future CO_2 emissions. A country can achieve carbon mitigation credits by reducing domestic emissions through the introduction of improved technologies or by financing initiatives that promote forest conservation, regeneration or establishment of artificial (plantation) forests.

According to the IPCC, deforestation caused the release of 1.6 billion tons of carbon (1 ton of carbon = 3.67 tons of CO_2) into the atmosphere. The panel estimates that 450 million hectares of forest have to be planted to sequester the 2.9 billion tons of carbon accumulated in the atmosphere as a result of all past emissions. Joint implementation can be an important instrument for increasing forest areas and sequestering the CO_2 currently in the atmosphere or at least slowing deforestation and land use change to avoid additional emissions.

Ecosystem studies conclude that dry biomass in tropical forests varies between 150 and 382 tons/ha. Assuming a carbon-to-dry biomass ratio of 45 per cent (Houghton *et al.*, 1991; Brown and Adger, 1993), their long-term carbon storage capacity could be between 67.5 and 171 tons/ha. Fearnside and Malheiros (1996) measured biomass levels of 52.8 tons/ha in 5-year-old stands of Brazilian secondary forest and levels of 196.6 tons/ha in 25-year-old stands.

Andersen (1996) estimated long-term carbon storage capacity at 140 tons/ha in mature primary forests, 55 tons/ha in partially intervened forests, and 10 tons/ha in pastureland. Calculations for primary forests in Costa Rica indicate biomass levels between 167 and 283 tons/ha and between 152 and 237 tons/ha for secondary forests (Carranza, 1996). This is equivalent to 75.15 to127.3 and 68.4 to106.6 tons of carbon per hectare, respectively.

These figures indicate that both artificial (plantation) and natural (primary and secondary) tropical forests might be important carbon sinks. Lack of information about the economic value of the environmental services provided by forests and of appropriate compensation to forest owners for these services increases the vulnerability of these ecosystems to land use change. Therefore it is important to determine their carbon sinking capacity and to estimate the economic value of this critical environmental service.

The first section of this chapter will review the different types of carbon sink services provided by forests and address the issue of placing economic values on them. The concept of joint implementation will be discussed in more detail and an example of its application will be given in the second section. The third section will present and illustrate various methods to estimate the magnitude and economic value of the carbon sink services rendered by each of the three main types of tropical forests: natural primary, naturally regenerating secondary, and artificial plantations. Alternative methods are needed due to data limitations or because of differences between the three forest types. The results are summarized in the final section.

VALUING OF CARBON SINK SERVICES

When valuing carbon sink services, it is important to distinguish between the three different accounting philosophies, carbon storage, carbon parking and carbon sequestration, reflected in current joint implementation programmes. Carbon storage relates to the capacity of a forest to maintain a certain amount of biomass per hectare. Retention of biomass means that the carbon in a forest is not being released into the atmosphere. In this case, pricing refers to a one-time payment for forest conservation. Land use change is permanently voided through, for example, the establishing of a national park. The value of the carbon storage service lies in avoiding potential future CO_2 emissions forever.

Carbon parking is less restrictive than carbon storage. Here pricing refers to the principle applied in the recent agreements with Costa Rican landowners. The agreements sanction land use changes during a limited period of time (20 years) in return for an economic compensation of US$200 per ha/year during the first five years of the agreement. Deforestation and land use change are at least forestalled, avoiding carbon emissions into the atmosphere while the agreement is in force. Carbon sequestration refers to the removal of CO_2 currently in the atmosphere, that is the mitigation of past emissions. Carbon sequestration by forests is a function of biomass growth rates. Payments are often made on an annual basis as long as the forest is a net absorber of CO_2.

Marginal social damage costs refer to the dollar value of the damage caused by the emission of an additional ton of carbon into the atmosphere (given the current levels), which is assumed to accrue to the global society as a whole. Therefore valuing carbon sink services at marginal social damage cost estimates is only conceptually correct in the case of carbon sequestration. Table 5.1 illustrates the wide range of marginal social damage costs estimated by various authors. The average of all available estimates is approximately US$20 per ton.

Table 5.1 Estimates of marginal social costs of CO_2 emissions (1990 US$)

Author(s)	Type of study	1991–2000	2001–2010
Nordhaus	MC	7.3	
Ayres and Walter	MC	30–35	—
Cline	CBA	5.8–124	7.6–154
Peck and Teisberg	CBA	10–12	12–14
Fankhauser	MC	6.2–45.2	7.4–52.9
Maddison	CBA/MC	5.9–6.1	8.1–8.4

Note: MC = marginal social cost; CBA = shadow value in a cost–benefit analysis.

Source: Fankhauser and Tol (1995).

Assuming that no country has the right freely to emit carbon on the basis of below-average total past emissions per person or unit-area, carbon storage and carbon parking services should command lower cross-country compensation levels, if any. Payments for carbon storage or parking within a country could be justifiable to achieve a more favourable national carbon emission balance. Nevertheless, developing country governments have been able to negotiate limited international compensation for carbon parking services. For example, Costa Rica signed a joint implementation agreement with the Norwegian government, receiving US$10 per ton of carbon parked in a specific forest area during a 20-year period. It has also issued 20-year carbon parking certificates that are being marketed internationally at the Chicago Mercantile Exchange (Costa Rican Office for Joint Implementation, 1995). Internally Costa Rica's law, through the National Investment Fund for Forest Financing, provides for similar levels of compensation to landowners for both carbon sequestration and carbon parking services (Decree No.26967, Ministry of Natural Resources and Environment (MINAE), 1996).

JOINT IMPLEMENTATION

Currently, the main vehicles for joint implementation are the Certificates of Tradable Carbon Offsets (CTOs), a concept supported by the World Bank and the Earth Council. These 'carbon certificates' embody the contribution of polluting countries towards solving the global warming problem through international initiatives that reduce CO_2 emissions to a specified amount. Joint implementation efforts render different benefits at the national and global levels (OCIC, 1995).

Benefits at the global level include:

- mitigation of global warming and the resulting adverse climate changes through the sequestration of atmospheric CO_2 and the reduction of potential future CO_2 emissions;
- implementation of environmentally sound and sustainable development initiatives in developing countries through the transfer of resources from industrialized countries.

Benefits for the country investing in carbon certificates include:

- ability to comply with international agreements to reduce greenhouse gas emissions when domestic mitigation measures are not sufficient or economically prohibitive;
- cost-effectiveness through the access to cheaper mitigation alternatives per ton of carbon sequestered or not emitted.

Benefits for the implementing country include:

- resources for investing in productive activities that prevent undesirable land use changes and promote biodiversity conservation, two fundamental pillars of sustainable development policies;
- the transfer of cleaner technologies for reducing national CO_2 emissions at a lower cost;
- generation of employment for forest owners and workers in rural areas and in urban areas through forest-related industries.

In tropical America, Costa Rica provides a good example of the potential of joint implementation. The Costa Rican Office for Joint Implementation (OCIC) was created in 1995 to seek and facilitate international JI agreements. At the local level, Costa Rica established the National Investment Fund for Forest Financing (FONAFIFO) to handle the payment for environmental services to farmers and forest owners (PSA) (Figure 5.1).

The OCIC is promoting Costa Rican projects that undertake a commitment to preserve natural forests, to promote or aid forest regeneration or to plant artificial forests (plantations) in return for a per hectare compensation by industrialized countries receiving equivalent carbon credits. For example, an agreement with Norway worth US$2 million was signed for the parking of 200 000 tons of carbon through watershed conservation and hydropower development in the Virilla Basin in the Central Valley (the price of US$10 per ton of carbon, lower than the average marginal damage costs figures mentioned above, could be attributed to the fact that this specific agreement was

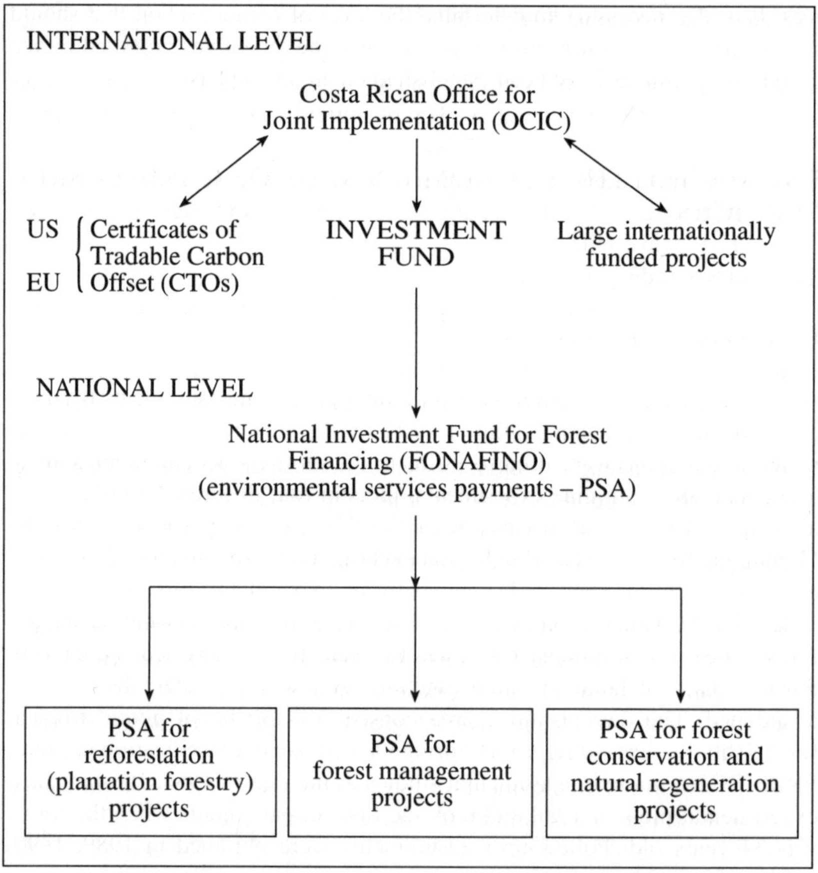

Note: US refers to the United States of America and EU to European Union countries.

Figure 5.1 Costa Rican apparatus for joint implementation initiatives

for carbon parking. Net sequestration was not required). Other joint implementation projects promoted by the OCIC include alternative energy generation and transportation solutions.

Costa Rica has also issued CTOs for 4 million tons of carbon and authorized the Financial Products Center Ltd. to sell them on the Chicago Mercantile Exchange. These have an estimated value of between US$40 and 80 million (OCIC, 1995). A more accurate estimation of the magnitude and economic value of the carbon sink services provided through plantation forestry and natural forest regeneration/conservation is needed because of the possibility of international initiatives that involve monetary compensation for these services. It is also necessary to determine the level of compensation that should be transferred to forest growers/owners, since the incentives provided by the current programmes have been established on an ad hoc basis.

ESTIMATING THE ECONOMIC VALUE OF CARBON SINK SERVICES

Tropical Secondary Forests

Data, methods and procedures

Data As a case study, seven years of individual tree measurements from the experimental and demonstration forests of the PROSIBONA silvicultural management project are used to estimate the biomass accumulation model. The data come from nine experimental plots in four different humid secondary tropical forest areas of Costa Rica: the 'Florencia' experimental area, the 'Tirimbina' forest, the Ian Hutchinson experimental farm, and the 'Espaveles' experimental farm. The stands range between two and 44 years of age.

The Ian D. Hutchinson demonstrative site is in a 44-year-old secondary forest. The original primary forest was first selectively harvested and then cut down to use the land for cattle ranching. A few years later, the site was abandoned. The demonstrative area consists of 180 ha of humid tropical forest (bh-T), according to Holdridge's life zone classification system. PROSIBONA started experimental studies at the site in 1987. It established experimental plots in 1988 and took the first measurements when the forest was 34 years old. Follow-up measurements were obtained in 1989, 1990, 1991, 1993, 1994 and 1995.

The 'Tirimbina' demonstrative site is in a life zone classified as very humid premontane forest transition to basal (bmh-P) and very humid tropical forest (bmh-T), according to Holdridge's life zone classification system. It has an extension of 29.16 ha and contains four experimental plots. Its origins are traced to the abandonment of rice crops after only one year of farming. Measurements were started in 1987 in two-, five-, 15- and 25-year-old forest plots. Follow-up measurements were taken annually until 1992 and then in 1995.

The 'Espaveles' demonstrative site is 20 ha of primary (13.4 ha) and secondary forest (6.6 ha) in a life zone classified as a humid premontane forest (bh-P), according to Holdridge. The area was deforested in 1937 and dedicated to rubber (*Hevea brasiliensis*) plantations in 1944. Other crops such as plantain and pineapple were grown later. Experimental studies on rice production were also conducted in the area, finally abandoned in 1954.

PROSIBONA took the first measurements in 1988 when the secondary forest was 35 years old. Follow-up measurements were taken annually until 1992 and then in 1997.

The 'Florencia' demonstrative site is 32 ha of secondary forest located in a very humid tropical forest (bmh-T) life zone, according to Holdridge. The soils in this area are believed to have a low productivity. The PROSIBONA project started measurements in 1993 when the forest was 27 years old. Follow-up measurements were taken in 1994, 1995 and 1998.

Biomass accumulation models Biomass accumulation is predicted for each individual tree using an equation estimated by Brown *et al.* (1989) with data from humid tropical forests:

$$Y = 13.675 - 6.1181D + 0.8391D^2 + e \tag{5.1}$$

where Y is total biomass in kg of dry weight, D is the tree's diameter at breast height (DBH), and e is an error term. Equation (5.1) has a coefficient of multiple determination (R^2) of 90 per cent. The biomass accumulation on each of the experimental plots at year-of-measurement t, which corresponds to a given forest age in that plot and site, is estimated by adding up the biomass predictions for all individual trees on the plot at year t and transformed into per hectare bases using the plot's area. The resulting data set consists of 50 biomass observations from forests of one to 44 years of age. This data set is used to estimate a non-linear model based on an adaptation of the forest volume growth function proposed by Richards (1959), which is widely used because of its flexibility. It can accommodate a wide variety of growth rates following near-linear, curvilinear and s-shaped patterns and always ends on a plateau. The adapted model is:

$$Y(t) = a_s(1 - exp^{-bt})^c + e; \; s = 1,2,3,4 \tag{5.2}$$

where $Y(t)$ is biomass accumulation as a function of forest age (t), a_s estimates the maximum biomass accumulation capacity in each forest site s (1=Florencia, 2=Tirimbina, 3=Ian Hutchinson, 4=Espaveles), b is the growth rate and determines the amount of time that it takes for the function to reach its maximum[2] (a_s), and c does not have a particular biological interpretation.

The error term is assumed to be an independently distributed normal random variable with zero mean and variances σs^2 (s=1,2,3,4) to account for differences in the variability in forest growth across sites. For model estimation, the Newton–Raphson algorithm in the OPTMUM procedure of the Gauss matrix algebra language programme is used to maximize the log-likelihood function given in equation (5.3),

$$LL = \sum_{t=1}^{50} -ln(\sigma s) - 0.5\{[Y(t) - a_s(a - exp^{-bt})^c]/\sigma s\}^2 \qquad (5.3)$$

where σs equals $\sigma 1$, $\sigma 2$, $\sigma 3$ or $\sigma 4$ and a_S equals a_1, a_2, a_3 or a_4, depending on whether the biomass observation corresponds to site 1, 2, 3 or 4.

Unfortunately, there are no data on the specific location of Costa Rica's secondary forest areas, which could be used to determine the corresponding life zones or to classify them by quality/productivity. Therefore a restricted model (a_1=a_2=a_3=a_4) is also estimated and used to forecast average biomass accumulation levels through time and maximum biomass accumulation capacity in all four sites for extrapolation to all Costa Rican secondary forests:

$$Y(t) = a(1 - exp^{-bt})^c + e; \sigma^2(e) = \sigma s^2(s = 1,2,3,4) \qquad (5.4)$$

The log-likelihood function to be maximized for estimating the model in equation (5.4) is obtained by setting a_1, a_2, a_3, and a_4 equal to a in equation (5.3). The main objective of the model is to predict average biomass accumulation with acceptable accuracy. Confidence bands using two alternative procedures are constructed to assess the reliability of the restricted model in predicting average forest biomass levels through time. The first procedure is based on Krinsky and Robb's (1986) Monte Carlo simulation technique to obtain confidence intervals for point estimates. This relies on the well-known result that maximum likelihood estimators are consistent, asymptotic and normally distributed, with covariance matrix given in equation (5.5),

$$CM = -\{E[I]\}^{-1} \qquad (5.5)$$

where E is the expected value operator and **I** is the matrix of second partial derivatives of the likelihood function (Judge *et al.*, 1985). The estimate of *CM* calculated by the Newton–Raphson algorithm is used to simulate 2500 sets of values of the coefficients of equation (5.4), as follows:

$$\mathbf{SV} = \mathbf{BML} + N * CHOL(\mathbf{CM3}), \qquad (5.6)$$

where **SV** is a 2500 by three matrix of simulated coefficient values, **BML** is a 2500 by three matrix with repeated rows containing the maximum likelihood estimates for a, b, and c in equation (5.4), **CM3** is the corresponding covariance matrix (given by the 3×3 principal minor of the estimated covariance matrix for the full model *CM*), N is a 2500 by three matrix of independently distributed standard normal random variables, * is a matrix multiplication operator and *CHOL*(**CM3**) indicates the Cholesky decomposition of **CM3**.

Equation (5.6) simulates 2500 sets of coefficient values such that *E*[**SV**] = **BML** and *Var*[*SV*] = **CM**3. In other words, it simulates the empirical joint probability distribution of the maximum likelihood estimators of the model coefficients (Ramirez, 1997). Let the four columns of **SV** that contain the 2500 simulated values of each coefficient be denoted by *SVa*, *SVb*, *SVc* and *SVs*. Then equation (5.7) provides 2500 simulated biomass levels at time t:

$$SY(t) = SVa(1 - exp^{-SVbt})^{SVc} \tag{5.7}$$

This is the empirical probability distribution of the model's prediction of average biomass accumulation at time *t*. The boundaries of a 90 per cent confidence interval for the prediction, therefore, are obtained by finding the value of the 125th (0.05 × 2500) and the 2375th (0.95 × 2500) largest simulated biomass level. The process is repeated for t=1,…,40 to build a 90 per cent confidence band for the model's biomass predictions though time.

The second procedure is based on Goldberg's (1964) approximation for the variance of a non-linear function of a set of random variables with known means and covariance matrix. Here the non-linear function is $Y(t) = a(1\text{-}exp^{-bt})^c$, the random variables are the estimators for *a*, *b* and *c*, their expected values *a*, *b* and *c* are the corresponding maximum likelihood estimates, and their covariance matrix **CM**3. Then:

$$\begin{aligned} Var[Y(t)] \approx {} & [\partial Y(t)/\partial a \; \partial Y(t)/\partial b \; \partial Y(t)/\partial c] * \mathbf{CM}3 * \\ & * [\partial Y(t)/\partial a \; \partial Y(t)/\partial b \; \partial Y(t)/\partial c]' \end{aligned} \tag{5.8}$$

where ∂ denotes a partial derivative, and the partial derivatives of *Y*(*t*) are evaluated at the maximum likelihood estimates for *a*, *b* and *c*. A two-standard error confidence band is given by:

$$\hat{C}B(t) = \Pi(t) \pm 2\sqrt{Var[Y(t)]} \tag{5.9}$$

where $\prod(t)$ is the average biomass accumulation level predicted by the model at time *t*.

This second approach has two disadvantages. First, it is based on an upper-bound approximation of the true variance of the average biomass prediction (that is, for the true population values of *a*, *b*, *c*, and **CM**3, equation (5.8) overestimates the variance). Second, the probability that the true average biomass accumulation level at forest age *t* falls within *CB*(*t*) is unknown, since *Y*(*t*) is likely not normally distributed. However, confidence-band estimates based on two different procedures are useful for a more rigorous assessment of the accuracy of the model's predictions, which is critical for a reliable economic analysis. A factor of 0.45 is used to transform the estimated biomass predictions to carbon storage (Houghton *et al.*, 1991; Brown and Adger, 1993).

Valuing of carbon sink services If sequestration were the prime objective of joint implementation agreements, payments should mostly be made for regenerating forest areas during their first years of significant growth. The total payment in this case should be based on the full mitigation or marginal social damage cost. Additional payments could be made for the conservation of older secondary forest areas to avoid land use change and the resulting CO_2 emissions. The average of the available estimates of the marginal social damage cost of CO_2 emissions (US$20 per ton of carbon) is used in this case.

Due to the lack of detailed forest inventory data, five informed assumptions are necessary to extrapolate per hectare values to all of Costa Rica's humid secondary forests:

1. The best secondary forest cover estimates available are correct: 425 000 ha, of which 302 000 (71 per cent) are humid. Of the latter, 160 000 (53 per cent) are less than 15 years old and the remaining are more than 15 years old (Solórzano *et al.*, 1991).
2. Significant secondary forest growth started in the late 1940s (Solórzano *et al.*, 1991).
3. Within the 0–15 and the 16–50-year-old categories, the forest age distribution is homogeneous.
4. Secondary forest areas will continue to grow at the estimated 1984–94 rate of 18 000 ha/year (Kaimowitz, 1995; CORFOGA, 1995) for 10 more years and at half that rate thereafter.
5. Of the future growth in forest areas, 71 per cent will be humid secondary forests; that is, the current humid-to-dry secondary forest proportion will be maintained.

It is also assumed that every year there is a choice between a one-time compensation of US$20 per ton sequestered and a payment of US$0.50/year per ton parked. The estimated model is first used to determine at what forest age (t^*) it is optimal to switch from requesting compensation for carbon sequestration to carbon parking, assuming the former payment rates. Given a strictly concave biomass accumulation curve (Figure 5.4), per hectare compensation for carbon sequestration is a monotonically decreasing function of forest age, while per hectare compensation for carbon parking is a monotonically increasing function of forest age. Therefore the optimal t (t^*) is found by solving:

$$0.5Y(t) = 20\{dY(t)/dt\}; \text{ or}$$
$$0.5\{a(1 - exp^{-bt})^c\} = 20\{cab(exp^{-bt})(1 - exp^{-bt})^{c-1}\} \quad (5.10)$$

Then let **M** be a 100×50 matrix containing the assumed age structure of Costa Rica's humid tropical forest areas, **F** be a 100×1 vector containing zeros in the first t^* rows and the per hectare biomass accumulation levels predicted by the model (*Yt*) for forests from t^*+1 to 100 years of age in the last 100-t^* rows, and **dF** be a 100×1 vector containing the yearly increments in per hectare forest biomass predicted by the model (*dYt*/*dt*) for forests between one and t^* years old in the first t^* rows and zeros thereafter. The flow of payments (*PFP*) accruing to Costa Rica's humid secondary forests for their carbon sink services during the next 50 years is estimated by the 50×1 vector:

$$PFP = S + P = \mathbf{M}'(0.45 \times 0.5 \times \mathbf{F}) + \mathbf{M}'(0.45 \times 20 \times dF) \quad (5.11)$$

where *S* is the flow of payment for carbon sequestration and *P* that for carbon parking. To build pessimistic and optimistic scenarios of the evolution of *S*, *P* and *PFP*, the elements of **F** and **dF** are obtained from the boundaries of the 90 per cent confidence band for the average biomass accumulation through time instead.

Results

Biomass accumulation Biomass accumulation models are estimated using a combination of time series and cross-sectional data from four forest sites representative of three different ecological areas of Costa Rica.

Figures 5.2 and 5.3 provide an overview of the biomass data calculated using measurements of all individual trees on each of the one hectare experimental plots and equation (5.1). The irregularities in biomass accumulation levels through time are due to the relatively small size of the measurement plots. It seems that biomass can reach anywhere between 100 and 200 tons per ha in 25- to 40-year-old stands. Secondary forest biomass at the various sites seems to reach a plateau at different levels, but at about the same age (30–35 years).

Table 5.2 presents the results from the estimation of the unrestricted and restricted biomass accumulation models. The full model (equation 5.2) has a coefficient of multiple determination (R^2) of 62.3 per cent, which is excellent for a model estimated combining times series and cross-sectional data (Judge *et al.*, 1987). The concentrated log-likelihood function reaches a maximum value of –166.44. The relatively small asymptotic standard errors of the parameters that estimate maximum biomass accumulation (a_1, a_2, a_3 and a_4) suggest that the secondary forests at the four sites reach different maximum levels: 101.54 tons/ha in Florencia, 190.11 tons/ha in Tirimbina, 142.53 tons/

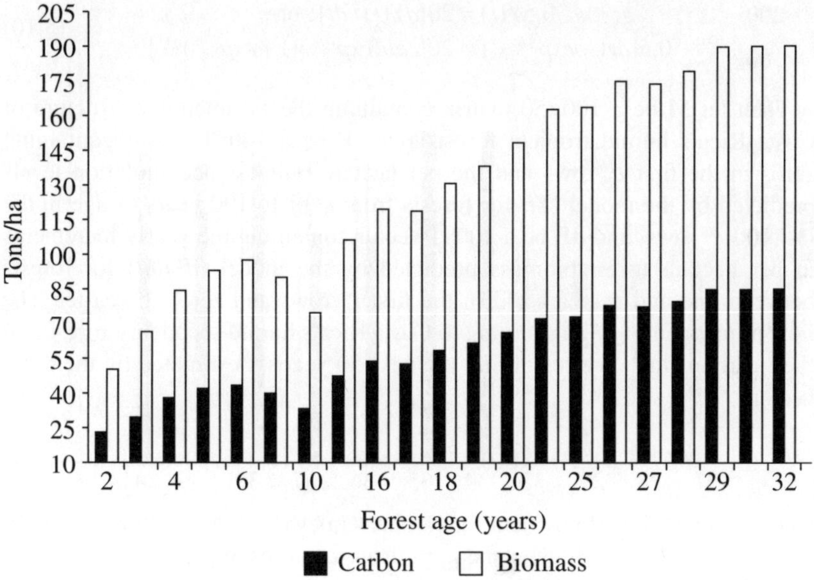

Figure 5.2 Biomass and carbon accumulation in a young Costa Rican tropical humid secondary forest (site 1)

ha in Ian Hutchinson, and 162.13 tons/ha in Espaveles (Table 5.2 and Figure 5.4). These differences are attributed to site quality.

Individual $\chi^2_{(1)}$ test statistics (not presented) indicate that none of the a_is are equal to one another ($a_1 \neq a_2 \neq a_3 \neq a_4$). Biomass variability appears similar in sites 2 and 3. Restricted model II imposes that additional assumption (*Ho*: σ2=σ3), which is not rejected by the likelihood ratio test (Table 5.2). Statistically, restricted models I and II are almost identical. The further restriction *Ho*:σ1=σ2=σ3=σ4, imposed in model III, is strongly rejected by the likelihood ratio test. Also individual $\chi^2_{(1)}$ test statistics (not presented) suggest that σ1≠σ2=σ3, σ1≠σ4 and σ4≠σ2=σ3.

Maximum biomass accumulation levels are statistically different across sites, but, because of the previously discussed aggregate data limitations, the four-site average estimate of a=144.46 tons of biomass (65.00 tons of carbon) per hectare from restricted model II will be used for countrywide projections. A key advantage is that the standard error associated with that estimate is only 2.38 tons of biomass (1.07 tons of carbon) per hectare (Table 5.2). The standard errors associated with the estimates of the other two coefficients (*b* and *c*) that determine the shape of the forest growth function before its maximum are also relatively low in restricted model II.

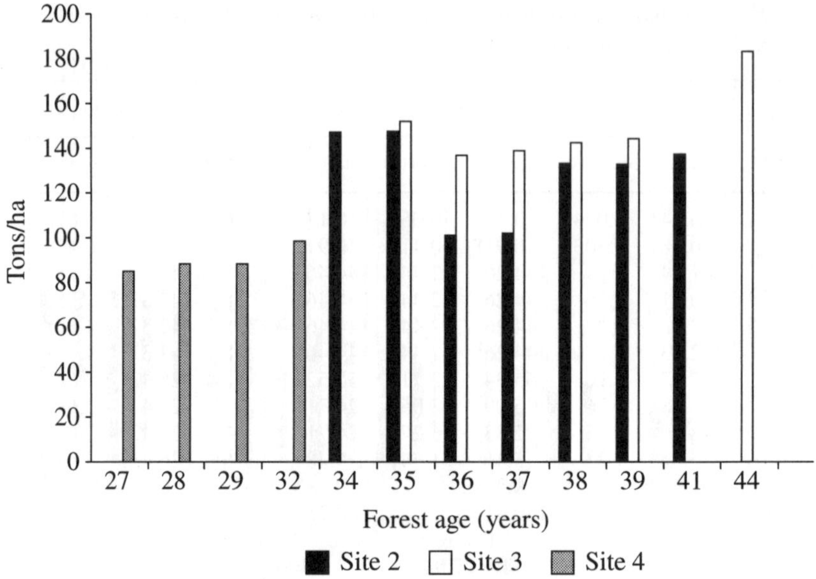

Figure 5.3 Biomass accumulation in three Costa Rican tropical humid secondary forests between 27 and 44 years of age

The restricted model forecasts that 97 per cent of the maximum biomass level is reached in 20 years, while nearly 85 per cent occurs during the first 10 years of forest growth (Figure 5.4). Therefore this ecosystem sequesters an average of 5.4 tons of carbon per hectare per year during the first 10 years of growth, and nearly an additional one ton/ha/year during the next 10 years. Thereafter, it can permanently store or park 65 tons of carbon per hectare. In comparison, preliminary estimates for artificial tropical forestry plantations indicated a carbon sequestration capacity of 7.7 tons/year during the first 15–20 years of growth (Ramirez and Gómez, 1999). However, the establishment costs can exceed US$1000/ha. Mature tropical primary forests do not sequester significant amounts of carbon, but can store between 80 and 90 tons/ha (Ramirez *et al.*, 2000).

Figure 5.5 presents the confidence bands calculated for the average biomass accumulation predictions from restricted model II, according to the two alternative methods described above. Both bands are wider at the beginning and become tighter with forest age. This is due to the more precise estimate obtained for the parameter that determines the maximum average biomass level attained by these forests. The estimates for the coefficients b and c, which control the shape of the function before its maximum, are less precise (Table 5.2). As expected, the two standard error confidence band based on

Table 5.2 Estimation results for the unrestricted and restricted biomass accumulation models

	Full model		Rest. mod. I		Rest. mod. II		Rest. mod. III	
Parameter	Est.	Std error	Est.	Std error	Est.	Std error	Est.	Std error
B	0.041	0.024	0.174	0.064	0.172	0.062	0.181	0.076
C	0.481	0.303	0.918	0.417	0.912	0.408	0.930	0.470
a1	101.54	2.29	144.26	2.44	144.46	2.38	142.37	4.75
a2	190.11	1.36	144.26	2.44	144.46	2.38	142.37	4.75
a3	142.53	1.37	144.26	2.44	144.46	2.38	142.37	4.75
a4	162.13	1.77	144.26	2.44	144.46	2.38	142.37	4.75
σ1	2.17	1.76	55.34	22.71	55.52	22.78	26.42	2.64
σ2	23.18	1.26	25.09	3.05	24.71	2.71	26.42	2.64
σ3	18.10	1.33	22.83	6.27	24.71	2.71	26.42	2.64
σ4	5.85	1.33	5.46	1.74	5.48	1.76	26.42	2.64
R^2	0.623		0.380		0.380		0.382	
MVLF	166.44		180.21		180.25		188.70	
Likelihood ratio test statistics			Full v. rest. I = 27.54		Rest. I v. rest. II = 0.08		Rest. II v. rest. III = 16.90	
Level of significance			1%		Insignificant at 10%		1%	

Notes: All parameter estimates in the restricted models are statistically different from zero at the 1 per cent level; MVLF refers to the maximum value reached by the concentrated log-likelihood function.

Goldberg's (1964) approximation is wider than the 90 per cent confidence band constructed using Krinsky and Robb's (1986) Monte Carlo simulation method, since the former relies on an upper-bound approximation of the variance of the prediction.

Another difference between the alternative confidence bands is that the simulated band reflects the non-normality of the average prediction errors. This is expected, since the average predictions are based on a non-linear function of normally distributed parameter estimator distributions. Notice, for example, that at forest age 10, Goldberg's band symmetrically deviates 18.6 tons/ha from the mean prediction of 120.5 tons/ha, while Krinsky and Robb's stands at 11 tons/ha above and 16 tons/ha below the mean prediction. To be safe, Goldberg's confidence band will be used for assessing the reliability of the economic-value estimations, since it is always at least as wide as Krinsky and Robb's.

Valuation of carbon sink services The optimal switching time (t^*) is calculated using the estimates of a, b and c (Table 5.2) and equation (5.9). Per

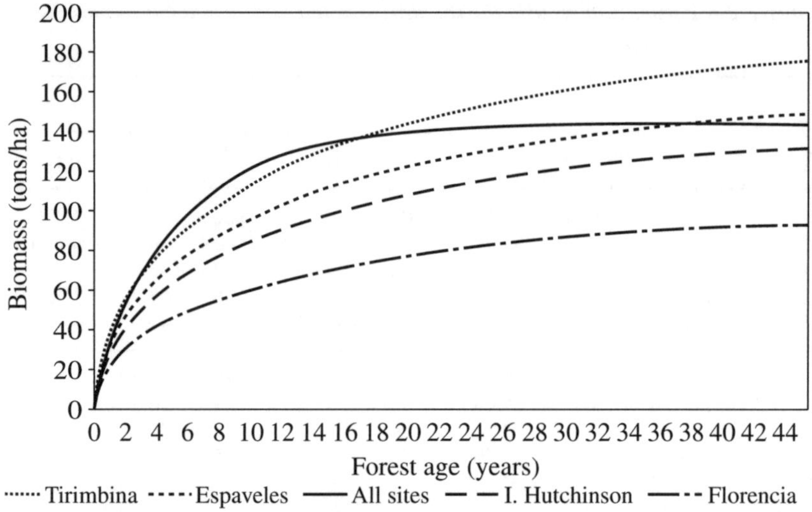

Figure 5.4 Biomass accumulation according to the restricted and unrestricted models

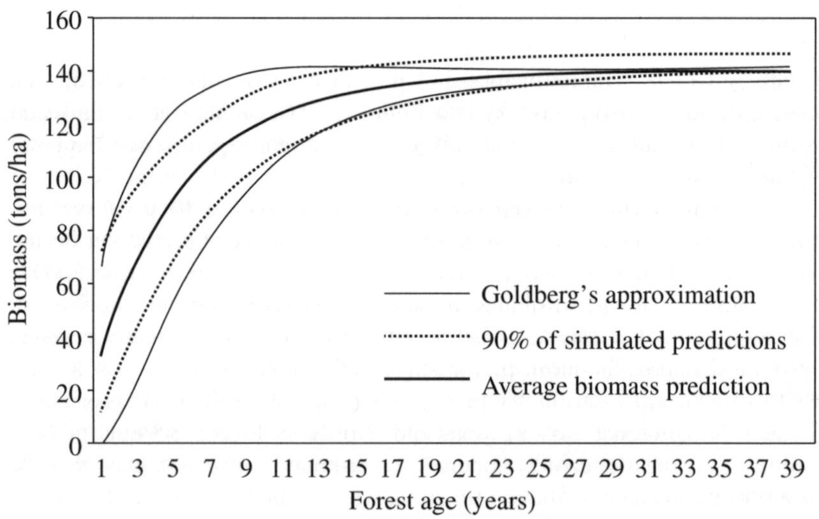

Figure 5.5 Alternative confidence bands for the average biomass predictions

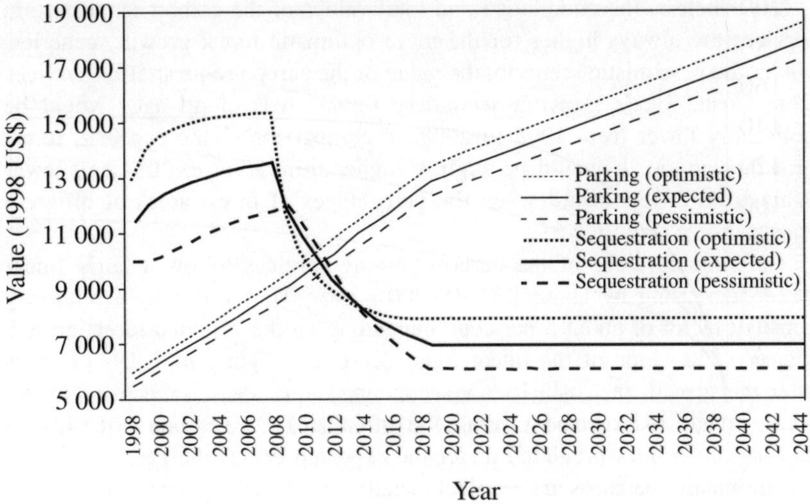

Figure 5.6 Estimated annual value of carbon parking and sequestration services from Costa Rica's secondary forests

hectare revenues from the carbon sink services are maximized by requesting annual payments for sequestration until the forests are $t^{*}=11$ years old and asking for carbon parking compensation thereafter. The expected evolution of the two components of PFP (equation (5.10)) (that is, the value of payments for carbon sequestration (S) and for carbon parking (P) during the next 50 years and their evolution under the previously discussed pessimistic and optimistic scenarios) is presented in Figure 5.6. The difference in the annual values of the carbon sequestration service under the expected and the pessimistic and the optimistic forest growth scenarios is wider at the beginning, reflecting that most of the sequestration occurring from 1998 to 2008 comes from young secondary forests (recall that the confidence band on per hectare biomass accumulation is much wider at earlier forest ages). From 2009 to 2014, the annual carbon sequestration values under the pessimistic scenario are higher than those under the expected scenario, which, in turn, are higher than those under the optimistic forest growth scenario.

During the latter period, the higher percentage of forest areas experiencing growth is between 6 and 11 years old. Notice in Figure 5.5 that, at those forest ages, the slope of Goldberg's lower-bound curve is higher than the slope of the average prediction function, which is higher than the slope of the upper-bound curve. Since the annual value of the carbon sequestration services is based upon yearly biomass growth, it is likely that forest growth paths that are slower overall produce higher annual values during certain periods.

Nevertheless, the cumulative and total values of the carbon sequestration services are always higher for the more optimistic forest growth scenarios. Under the pessimistic scenario, the value of the carbon sequestration services from Costa Rica's growing secondary forest areas of all ages would be noticeably lower from 1998 to 2008, in comparison to the expected forest growth scenario. It would be slightly higher from 2009 to 2014 and lower again from 2018 onward when the percentages of forest areas of different ages stabilize (Figure 5.6).

The annual values of the carbon parking services follow a fairly linear trend, increasing by about US$400 000 a year. They differ by a relatively constant factor of about 5 per cent, depending on the assumed forest growth scenario The slope of the linear trend decreases slightly in 2019, 11 years after the growth in Costa Rica's secondary forest areas is assumed to decrease by half, as expected. Overall, the differences in the carbon sequestration and parking values predicted under the expected versus the pessimistic and the optimistic scenarios are relatively small.

In 1998, Costa Rica's forestry sector could claim a compensation of US$11.9 million ($10.3 to $13.4 million according to the pessimistic and optimistic forest growth scenarios, respectively) for carbon sequestration and an additional US$5.7 ($5.45 to $6) million for carbon parking in forests over 11 years of age. As older forest areas accumulate, the potential revenues from carbon parking services increase to US$10.4 ($9.8 to $11.0) million in 2011, when they equal the estimated potential revenues from carbon sequestration.

The latter decline after 2008 because it is assumed that humid tropical secondary forest areas only increase at half the previous rate (that is, by 6400 ha/year) beyond that year. By 2048, the potential total revenues from the carbon sink services provided by a still-modest forest cover of 680 000 ha (about twice today's areas), are estimated at US$26.4 ($28.0 to 24.8) million per year. A much more aggressive growth in secondary forest areas would be necessary to compensate for the deforestation that occurred in Costa Rica just during the 1980s. Nevertheless, the potential total revenue from carbon sequestration and parking during the next 50 years is estimated at 1115 (1038 to 1194) million 1998 US$ (669 (642 to 697) for parking and 446 (396 to 497) for sequestration), with a net present value of US$740 (687 to 794) million (436 (418 to 454) for parking and 304 (269 to 340) for sequestration) using a real social discount rate of 3 per cent per year. This is nearly 10 per cent of Costa Rica's 1998 gross domestic product.

Given the assumed compensation rates for carbon sequestration and parking, the 100-year value of the carbon sink service of a single hectare of humid tropical secondary forest under the expected average growth scenario is US$3876 ($3683 and $4218 under the pessimistic and optimistic scenarios, respectively). The corresponding net present value is of US$1689 ($1506 and

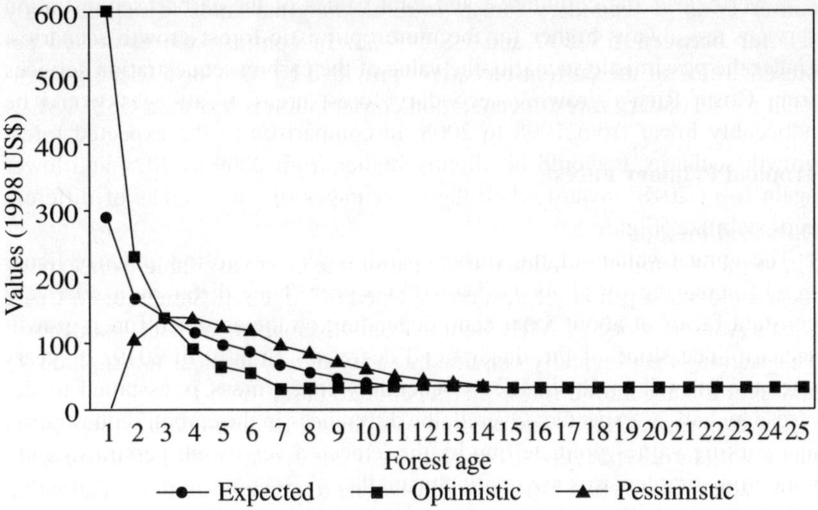

Figure 5.7 Estimated annual values of the carbon sink services of an average hectare of Costa Rican secondary forest

$1948) at a real social discount rate of 3 per cent and US$1002.5 ($796 and $1254) at a real private discount rate of 8 per cent.

In the single-hectare case, most of the compensation accrues at the beginning, due to the substantial carbon sequestration that occurs during the first 11 years of aggressive forest growth (Figure 5.7). Under the optimistic scenario, most of the value is accrued during the first four years of very rapid forest growth and, under the expected scenario, during the first six years. The pessimistic scenario accumulates most of the value from the second to the eighth year of forest growth in a more levelled fashion. The flow of annual values looks very much the same for the three scenarios after the twelfth year of forest growth.

An average hectare of regenerating humid tropical secondary forest should entitle a payment of US$1120 for sequestering 56 tons of carbon during the first 11 years of forest growth. It could also be subject to an additional incentive of US$640 for every additional 20-year period that land use change is avoided and the sequestered carbon remains parked. If inflation-adjusted, evenly distributed payments are made every period for sequestration and later for parking, assuming a real discount rate of 8 per cent, this stream of income has a net present value of approximately US$1000 per ha, about US$830 (83 per cent) for sequestration and US$170 (17 per cent) for parking.

This would be a strong incentive for a typical landowner in Costa Rica and in most other developing countries to permit the regeneration and assist in the conservation of secondary forests in rural, marginal lands, which currently sell for between US$600 and US$900/ha. In contrast, the estimated net present value of the current incentives provided by the Costa Rican government for secondary forest regeneration/conservation is less than US$200/ha.

Tropical Primary Forests

Data and methods

Six years of detailed individual tree measurements are available from experimental plots in a primary forest area located in Virgen de Sarapiqui on Costa Rica's Atlantic slope (CATIE Natural Forest Management Unit Data Base). This region is ecologically classified as very humid tropical forest (bmh-T) and 'premontane humid forest' in transition to basal (bmh-P).

The experiment has a nine random plot design accommodating three treatments with three replications. The plot size is one hectare with a 40 metre-wide buffer zone (Quiros and Finnegan, 1994). The treatments are three alternative silvicultural models: conventional harvesting (control), conventional harvesting with a posterior upper canopy treatment (upper canopy), and conventional harvesting followed by a liberation treatment (liberation) to promote the establishment of tree species with high commercial value. The conventional harvesting was done in 1989–90 and the silvicultural treatments were applied in 1991 and 1992 (Finnegan *et al.*, 1993).

Annual measurements have been taken from 1990 to the present. The six years of data from the experimental plots are used to simulate 54 additional years of forest growth using the Natural Forest Yield Simulation Program (SIRENA 2). The simulation accounts for the different silvicultural treatments applied and assumed conventional harvesting in years 20, 40 and 60. Individual tree surface biomass from year one to 60 is estimated for all trees on each of the treatments and replications using the actual and the simulated single-tree diameter data and two different regression equations fitted by Brown *et al.* (1989) and Brown (1997, personal communication) for tropical humid primary forests:

$$B = 13.2579 - 4.8945(D) + 0.6713(D^2) \tag{5.12}$$

$$B = 21.2970229 - 6.952649(D) + 0.740300(D^2) \tag{5.13}$$

where B is the prediction of the total dry, above-ground biomass in kilograms for an individual tree with a chest-height diameter D. The total surface biomass per hectare during year t (t=1,…60) for each of the treatments is estimated by adding up all of the single-tree biomass predictions in the three (one-hectare) replications and dividing by three. According to the Forest

Products Laboratory of the United States Department of Agriculture (USDA), the carbon content for most of the forest species found in Costa Rica is between 47 and 53 per cent. Given the total biomass per hectare, the carbon content is obtained multiplying by 0.50.

In this case, a formal financial analysis is conducted. The information about the direct cost and benefits of forest production is obtained from CATIE's Natural Forest Management Unit Data Base. The cost–benefit analysis method is first used to determine the financial profitability of each silvicultural treatment, ignoring the potential payments for the carbon sink services. The time value of money is taken into account by calculating net present values (NPV), using real discount rates of 5 per cent, 8 per cent, and 10 per cent.

The value of the carbon sink service is incorporated through an expanded financial analysis where the potential payments for this service are also considered as income. This is only a partial environmental–economic analysis since it does not include the value of other environmental services from forests such as biodiversity conservation, soil protection, water regulation, scenic beauty, recreation and so on. All costs and benefits are converted to US dollars using the official exchange rate posted by the Bank of Costa Rica. Since mature primary forests mostly store carbon, the value and, thus, the payment for the carbon sink service are assumed to be US$10 per ton parked for a 20-year period. This is consistent with the compensation rate agreed upon in the most recent joint implementation programme between Costa Rica and Norway.

The NPVs and the AC/B ratios are calculated for different scenarios. First, it is assumed that silvicultural treatments are carried out (where pertinent) and the forest is harvested 20 years later. This is a monoperiodic or short-term evaluation of the forest management activity. Second, it is assumed that the forest is harvested in year one, the silvicultural treatments (where pertinent) take place in year two, and periodic harvests are carried out every 20 years, ad infinitum. This is a multi-period or long-term evaluation of the forest management activity. For each multi-period analysis (financial and expanded financial), the NPV is calculated for the first three cycles (NPV1=2–20 years, NPV2=21–40 years and NPV3=41–60 years). It is assumed that the NPV for the later cycles is the same as NPV3. Then, by the Fausstmann formula:

$$LNPV = VIH + NPV_1 + NPV_2 /(1+i)^T + NPV_3 /\{(1+i)^{2T}(1-e^{-iT})\}; \quad (5.14)$$

where:

LNPV = net present value of a long-term forest management project,
VIH = value of the initial harvest,

NPV1 = NPV from year 2 to 20, including the cost of the second-year silvicultural treatment and the cost and income from a harvest in year 20,
NPV2 = NPV from year 21 to 40, including the cost and income from a harvest in year 40,
NPV3 = NPV from year 41 to 60, including the cost and income from a harvest in year 60,
i = real discount rate,
T = length of the rotation (20 years).

Notice that in equation (5.14), NPV2 is discounted T=20 years by dividing it by $(1+i)^T$. The present value of an infinite periodic flow of income every 20 years in the magnitude of NPV3 is calculated by dividing NPV3 by $(1-e^{-iT})$ and discounted from year 2T=40 to year zero by dividing it by $(1+i)^{2T}$. For each treatment, equation (5.14) is used separately to calculate the long-term financial net present value (LFNPV), which does not include carbon payments and the present value of a long-term stream of income from the carbon storage payments (LNPVC). The sum of these two (LFNPVC) is the present value of the long-term net income per hectare accruing to an owner for managing the forest in a sustainable manner and participating in the carbon storage programme.

Results

Very similar biomass estimates are obtained with equations (5.12) and (5.13). The results from equation (5.12) are used for the analysis. The liberation treatment plots had a higher initial average biomass level of 174.79 tons/ha v. 158.35 in the control and 152.08 in the upper canopy treatment (Figure 5.8). The first harvest (year 1) reduced biomass in similar quantities, to 160.98, 142.56 and 140.04 tons/ha, respectively. The control began biomass recovery immediately in year 2, while the treatments caused additional biomass reductions to 121.44 tons/ha (year 4) in the case of liberation and to 138.22 tons/ha (year 3) in upper canopy.

Twenty years after the first harvest, the control and the upper canopy treatment achieved biomass levels higher than the initial levels (189.67 and 181.76 tons/ha). The difference between them at year 20 (6.27 tons/ha) is almost identical to the original difference (7.91 tons/ha). The liberation treatment is far behind (168.88 tons/ha) and still below its initial biomass level. Liberation, however, could have a financial advantage, since it yields a higher-valued second harvest (26.33 tons/ha v. 22.79 and 18.08 tons/ha in the control and in the upper canopy treatment, respectively). The differences between the biomass levels of the control and both treatments become steadily smaller over time. During the last rotation simulated (years 41–60), 20-year average

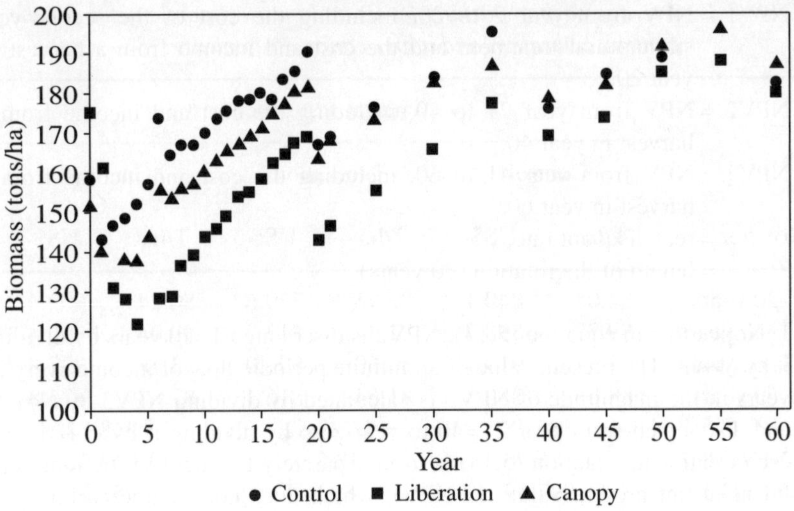

Figure 5.8 Biomass levels through time for the control, liberation and canopy treatments

biomass levels are closer (183.94, 179.2 and 187.55 tons/ha) and the final levels (year 60) are almost identical (Figure 5.8).

Table 5.3 shows the average carbon parking levels (that is, biomass × 0.50) for the control and the two treatments during the three 20-year rotations included in the analysis not adjusted and additively adjusted according to the original (year zero) carbon levels. Clearly, the liberation treatment is inferior to the control in this respect, while the canopy treatment is slightly better overall in the long term.

The economic value of the carbon parking service rendered during each rotation is also presented in Table 5.3. It is obtained by multiplying the average carbon parking levels by the assumed price of US$10/ton. The LNPVC (present value of the long-term carbon parking service; that is, carbon storage) is calculated using equation (5.14), taking into account the current scheme for environmental service payments (PSA) for natural forest conservation stipulated by the Costa Rican government. Decree 26976 establishes that 50 per cent of the payment is made 'up front', 20 per cent during the second year, and 10 per cent during each of the three following years of the 20-year contract.

The current amount (US$360) established by the Costa Rican Ministry of the Environment in decree no. 26976 as payment for environmental services rendered by sustainably managed natural forests represents less than half the economic value of the carbon parking service provided during the first 20-

Table 5.3 Average carbon parking levels and their economic value for the control, liberation and upper canopy treatments

	Without adjustment					
	Control		Liberation		Canopy	
Rotation	T/ha	US$	T/ha	US$	T/ha	US$
0–20 years	84.05	840.1	73.96	739.6	80.24	802.4
21–40 years	89.87	898.7	81.37	813.7	89.04	890.4
41–60 years	91.97	919.7	89.60	896.0	93.78	937.8
0–60 years	88.63	886.3	81.64	816.4	87.68	876.8
LNPVC 5%	1 385.48		1 247.62		1 348.72	
LNPVC 8%	1 085.96		965.86		1 047.52	
LNPVC 10%	997.19		883.45		958.72	
Adjusted to 80 tons of C/ha in year 0						
0–20 years	85.37	853.7	66.56	665.6	84.20	842
21–40 years	91.19	911.9	73.98	739.8	93.00	930.0
41–60 years	93.30	933.0	82.21	822.1	97.74	977.4
0–60 years	89.95	899.5	74.25	742.5	91.64	916.4
LNPVC 5%	1 407.04		1 129.16		1 412.14	
LNPVC 8%	1 103.16		871.74		1 097.90	
LNPVC 10%	1 013.09		796.57		1 005.21	

year rotation. Table 5.3 also shows that the liberation treatment has a long-term environmental cost in relation to the control. This is reflected by the differences in the corresponding LNPVC of between US$220 and US$280/ha (depending on the discount rate used) due to a lower average carbon parking rate over time. This cost is practically non-existent in the case of the canopy treatment.

In the more limited financial analysis, the income is the value of the wood products obtained from the first (year one) harvest and the later harvests simulated for years 20, 40 and 60 using SIRENA2. The average 1998 price per cubic metre of wood of different categories is converted to 1998 US dollars according to the official exchange rate of Costa Rica's Central Bank.

Four groups of species prevailed. Groups one and three include species classified as semi-hard and common semi-hard woods with average prices of 70.56 dollars/m^3. Group two includes species with higher commercial values, averaging 93.34 dollars/m^3. Group four consists of species utilized for ply-

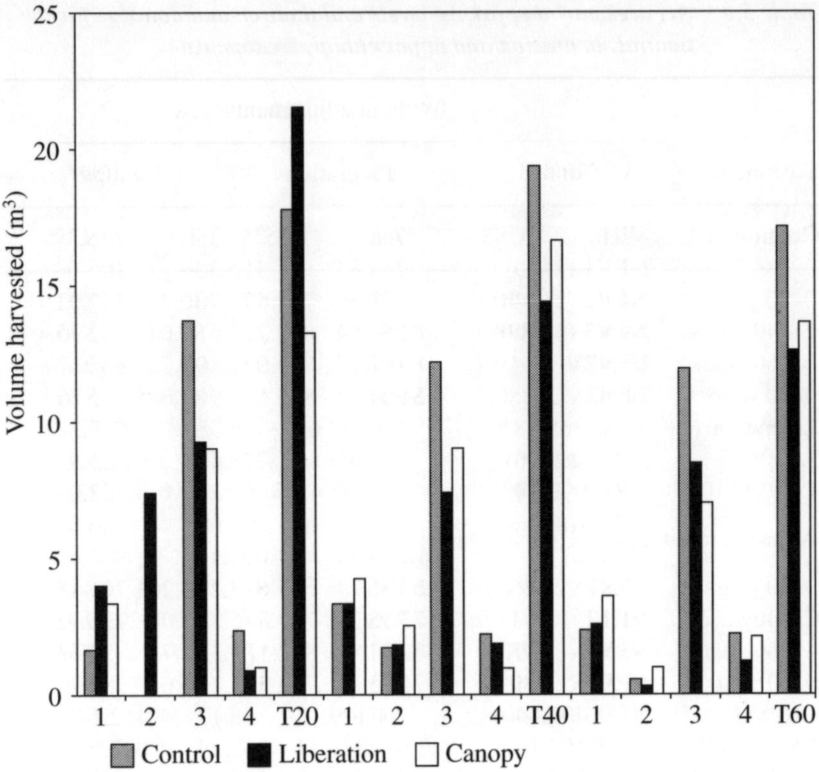

Figure 5.9 Volume harvested of four groups of species (1, 2, 3 and 4) and total (T20, T40 and T60) during the three harvesting cycles

wood that have a low commercial value of 39.57 dollars/m^3. Figure 5.9 illustrates the yields at the end of the first 20-year cycle. They are considerably higher in the liberation treatment, especially those of group two, which includes the most valuable species. However, the control treatment produced the best commercial yields in years 40 and 60, followed by the canopy treatment. For the three cycles together, the control produced 54.4 m^3/ha v. 48.6 and 43.7 for the liberation and canopy treatments, respectively. Excluding group three of low-valued species, the yields were 47.6, 44.6 and 39.6 m^3/ha for these three treatments, respectively.

Table 5.4 shows the monoperiodic net present value (NPV1) and the multiperiod or long-term financial and environmental–economic net present values (LFNPV and LFNPVC) for the control, liberation and canopy treatments under three different real discount rates, as well as, the NPVs for the second

Table 5.4 Net present value for the control, liberation and canopy treatments under three different real discount rates

Treatment	NPV	Discount rate 5%	8%	10%
Control	VIH	736.38	736.38	736.38
	NPV1 (2–20)	284.83	155.30	106.04
	NPV2 (21–40)	317.39	177.70	119.91
	NPV3 (41–60)	250.64	138.61	92.35
	LFNPV	1 197.15	937.80	862.60
	LFNPVC	2 604.19	2 040.96	1 875.69
Liberation	VIH	736.38	736.38	736.38
	NPV1 (2–20)	430.64	238.43	156.91
	NPV2 (21–40)	178.96	104.26	68.09
	NPV3 (41–60)	140.49	74.09	46.83
	LFNPV	1 266.03	1 001.46	904.61
	LFNPVC	2 395.19	1 873.20	1 701.18
Canopy	VIH	736.38	736.38	736.38
	NPV1 (2–20)	158.80	84.35	53.04
	NPV2 (21–40)	232.23	127.83	84.74
	NPV3 (41–60)	164.44	88.13	56.70
	LFNPV	1 019.66	853.24	803.47
	LFNPVC	2 431.80	1 951.14	1 808.68

and third production cycles (NPV2 and NPV3) that are necessary to calculate the former. Since the initial harvest in 1989/90 was done for the entire experimental area together (that is, plot-by-plot data were not taken), the value of the initial harvest (VIH) is similar for the control and the two treatments. It is the per hectare average for the whole experimental area.

Note that in the monoperiodic analysis (NPV1), the liberation treatment is far better than the control at any discount rate, while the canopy treatment is notably inferior. However, even at the low real discount rate of 5 per cent, the NPV1 of US$430.64/ha is little in comparison with the average value of a hectare of land used for forestry production in Costa Rica. Agricultural land is leased for US$74.14/ha/year, on average (Prices and Agroforestry Statistics Bulletin, 1993).

Assuming a similar real rate of return for land leasing (5 per cent), this would imply a per-hectare value of US$74.14/0.05 = US$1482.80. Forested land converted to ranching or agricultural uses should have a lower-than-average value, perhaps between 50 and 75 per cent or US$741.40–1112.1/ha. This range coincides with recent forest land price reports from various regions of the country.

Depending on the initial state of the forest, a first harvest can be done at the beginning of the sustainable management period, and can be included when doing the financial (that is private) or environmental–economic (that is, private plus environmental payment) evaluation. In this experiment, that first harvest earned US$736.38/ha (VIH in Table 5.4). If the value of the initial harvest is taken into account, both the control and the treatments show a monoperiodic net present value (VIH+NPV1) within or above the range of land values mentioned above, even at the highest real discount rate of 10 per cent. This makes them financially attractive, even from a short-term forest production perspective (that is, 20 years), compared with the alternative of selling the land.

The problem with this is that the initial harvest and the sustainable management necessary to produce after 20 years are two separable activities. The owner, if allowed, might decide to make a more intensive and lucrative harvest and then sell or convert the land to ranching or agricultural uses. Therefore it is important to evaluate if the sustainable management of natural forests is financially attractive without considering the value of a possible initial harvest. Given the high cost of land in Costa Rica, the results of this analysis indicate that sustainable forest management is not financially attractive in the short term, since the value of the land for alternative purposes is higher than NPV1. This might be different in other developing countries with lower land values.

The result of the long-term financial evaluation is similar, since the net present values of the harvests done 40 or more years in the future (VAN2 and VAN3) are relatively small. In the long run, landowners are likely to demand higher rates of return from their investments (that is, real discount rates) due to the additional risk associated with having to wait 40 or more years to recover part of their investment (which consists of the initial value or opportunity cost of the land and the operational costs for a sustainable forest management). In the case of long-term investments, it is more appropriate to assume a real discount rate of at least 8 per cent.

The liberation treatment also offers the best financial results in the long term, reaching a LFNPV of US$1001.46, v. US$937.80 for the control and US$853.24 for the canopy treatment. It is clear that the greater quantity of wood of high commercial value produced by the liberation plots in year 20 compensates for the lower yields in the later harvests, even though the differences in the long-term financial NPVs are small. These LFNPVs are within the range of land values mentioned above. However, they include the value of the initial harvest (VIH), which represents at least 75 per cent of them. If the

VIH is considered, the sustainable management of natural forests does not appear attractive in the long term in comparison with the value or opportunity cost of the land in Costa Rica. It might be financially feasible in other developing countries with land values of less than US$300/ha. The present value of the long-term environmental service of carbon storage (LFNPVC-LFNPV) is, depending on the treatment, nearly as high or higher than the net present value of the wood production activity (LFNPV).

The long-term environmental–economic net present values (LFNPVC) are US$2040.96, US$1873.20 and US$1951.14/ha ($i$=8 per cent) for the control, canopy and liberation treatments, respectively. These include both the private returns (that is, the financial value of the harvests) and the carbon-storage values from sustainable forest management. Even if the VIH of US$736.39/ha is excluded, the added payment for carbon storage services according to the mechanism and amounts specified previously is sufficient to make the sustainable management of natural forests attractive to the forest owners in relation to the alternative of selling or converting the land to other uses.

When the carbon storage values are taken into account, the control is the recommended treatment, while the great initial loss of biomass triggered by liberation (which only recovers very slowly) takes it from being the best treatment to having the lowest LFNPVC, 15 per cent less than the control. The control has slight LFNPVC advantage over the canopy treatment, which is merely financial, since the economic value of the carbon stored by both over time is similar.

Plantation Forestry

Data, methods and procedures

Tree cropping in the form of plantation forestry started in Costa Rica on a small scale more than 30 years ago. In 1964, 12 planted hectares were reported for the first time (Boletín Estadístico no.5) and small additional areas continued being registered annually until 1978. During this period, only 19 hectares per year were planted, on average. The increase in deforestation during those years started signalling an eventual depletion of the reservoirs of natural forest. It is estimated that, from 1963 to 1973, 48 800 hectares of forest per year were cut and 31 800 per year from 1973 to 1989 (Solórzano *et al*., 1991). The annual rate of deforestation in Costa Rica has ranged from 1.8 per cent in 1963 to 1.2 per cent in 1989, compared with the average rate for all tropical forests, which is estimated at 0.6 per cent per year.

Nearly 850 000 ha of forest were cut in Costa Rica between 1966 and 1984 of which only 35 per cent were turned into agricultural and livestock production. About 65 per cent of that area, on the other hand, was used for forestry production.

The search for alternatives to reduce the pressure on the remaining natural forests and prolong the useful life of the affected ecosystems resulted in an interest to promote forestry plantations through economic incentive programmes that started in 1978. In 1979, 413 hectares of trees were planted and, thereafter, an average of 2500 ha of additional plantations were established every year until 1989. In 1987, organized groups of small and medium-sized farmers joined this reforestation movement, motivated by incentive programmes that targeted such groups. They planted 34 640 hectares of trees between 1988 and 1994 (Torres et al., 1995); that is, almost 5000 annually. If the projects started by large forestry companies are taken into account, it is estimated that plantation forestry areas in Costa Rica have grown by an average of 16 759 hectares per year from 1990 to 1995 (Table 5.5).

Table 5.5 Hectares of plantation forestry established in Costa Rica, per year (1964–95)

Year	Area planted that year	Accumulated area planted	Year	Area planted that year	Accumulated area planted
1964–71		84	1984	1 286	6 222
1972	84	168	1985	2 501	8 723
1973	24	192	1986	4 175	12 898
1974	21	213	1987	5 303	18 201
1975	21	234	1988	4 835	23 036
1976	5	239	1989	5 000	28 036
1977	12	251	1990	13 797	41 833
1978	34	285	1991	15 560	57 393
1979	413	698	1992	15 958	73 351
1980	807	1 505	1993	14 630	87 981
1981	1 098	2 603	1994	14 628	102 609
1982	1 357	3 960	1995	25 981	128 590
1983	976	4 936			

Source: Ministerio de Recusos Naturales, Energía y Minas (1994b), *Boletín Estadístico Forestal*, no.5.

Between 1990 and 1993, the most commonly planted species was the 'melina' (*Gmelina arborea*) in about 44 per cent of the area established during that period. Other important species in this regard included 'pilón' (*Hieronyma alchorneoides*), 'eucalipto' (*Eucaliptus spp.)*, 'teca' (*Tectona grandis*) and 'laurel' (*Cordia alliadora*). Those five were planted in an area of 48 221 hectares, a full 80 per cent of the 1990–93 total.

Due to the limited amount of reliable data available on the long-term growth of tropical plantation forestry species, a simpler, less precise, procedure has to be used to estimate the absorption of CO_2 and the carbon storage occurring in Costa Rica's plantation forests.

1. Estimate the mean average annual increment (AAI) of trunk volume for the five most used species in Costa Rica, mentioned previously. The AAI was calculated for each of the five species using 10 years of annual growth field data (CATIE/Madeleña Project Silvicultural Guidebooks) representative of the conditions of forestry plantations in Costa Rica. The five-species mean AAI was then calculated at 18.9 cubic metres per hectare per year. The application of an adjustment factor of 0.85 to account for the previously mentioned deficiencies in the current status of the plantations (Torres *et al.*, 1995) results in a final mean AAI of 16.4 cubic metres per hectare per year.
2. Estimate the mean AAI of volume for the whole tree, including trunk, branches, leaves and roots. Schroeder *et al.* (1993) estimate that at least 40 per cent of a tree's biomass is in its branches and leaves. An evaluation of data from different types of tropical forests (Brown *et al.*, 1989) resulted in multipliers from 1.8 to 3.5. The values of this multiplier in other literature reviewed (Brown and Lugo, 1984; Thoranisorn *et al.*, 1991; Faeth *et al.*, 1994; Nabuurs and Mohren, 1994) range from 1.3 to 3.6, with an average of 2.5, for natural forests and between 1.3 and 2.7, with an average of 1.9, in the case of forestry plantations. A multiplier of 2.0 to convert trunk biomass to whole-tree biomass is used in this study, resulting in a mean AAI of volume for the whole tree of 32.8 m^3/ha/year.
3. Determine mean specific weight for the five principal species. Considering the known values of 0.37gr/cm^3 for 'melina' (*Gmelina arborea*), 0.70gr/cm^3 for 'teca' (*Tectona grandis*) and 0.50gr/cm^3 for 'eucalipto' (*Eucaliptus spp.*) (CATIE/Madeleña Project Silvicultural Guidebooks), their average of 0.47gr/cm^3 or ton/m^3 is used in this study.
4. Determine the carbon content of the wood. As stated above, according to the Laboratory of Forest Products of the USDA, the five principal species planted in Costa Rica contain between 47 and 53 per cent of carbon (Barres, 1993). Thus an average carbon content of 50 per cent was assumed.
5. Calculate the quantity of carbon stored per ton of wood biomass. The method described by Barres (1993) was followed for this purpose. The mean specific weight of Costa Rica's plantation forestry wood (0.47 ton/

m^3) is multiplied by its average content of organic carbon (0.50), yielding a conversion factor of 0.235.

6. Calculate the amount of carbon stored per hectare of plantation forestry per year. Multiplying the mean AAI of volume for the whole tree of 32.8 m^3/ha per year by the conversion factor in (5) above results in an estimate of 7.7 metric tons of carbon stored per ha/year.
7. Estimate the net additional quantity of carbon stored by forestry plantations in Costa Rica during each of the years of the study. For this, the accumulated number of hectares in the different years is multiplied by the previously calculated factor of 7.7 and the amount of carbon exiting the system every year through the harvesting of forest products is subtracted (Table 5.2, cols (1) and (2)). The harvested volumes reported by MINAE (1995) have to be multiplied by two, since they refer to trunk volumes only, and by the conversion factor given in (5) above.
8. Estimate the net accumulated quantity of carbon stored by forestry plantations in Costa Rica for each of the years of the study. This is done by accumulating through time the net additional quantity of carbon stored yearly, as calculated in (7) above.
9. Estimate average net quantity of carbon that has been stored in Costa Rica's forestry plantations, per year, during the last 25 years, by summing up the net accumulated quantity of carbon stored during each of the years of the study and divide it by 25. The equivalent amount of CO_2 sequestered can be calculated. It is known that 27.3 per cent of the CO_2 molecular weight is carbon. Thus each ton of carbon stored implies a sequestration of 3.66 tons of carbon dioxide.

Results

In column (1) of Table 5.6, the additional carbon stored during each year by the existing plantation forestry area is calculated, net of the amounts exiting the system in that time period (given in column (2)). The total amount of carbon that has been kept in storage during a given year is computed in column (3) by accumulating column (1) through time. Because of the larger areas planted in recent years and the fact that the carbon being stored during the previous periods has been accumulating, the net total amounts of carbon kept in storage increase rapidly at the end to 4.42 million metric tons by 1995.

On average, however, only 746 769 tons of carbon (per year) have been kept in storage in Costa Rica's forestry plantations from 1970 to 1995, while the average for the last 20 years rises to 932 563 metric tons. Obviously, if the 128 590 hectares already established remain in plantation forestry during the 20-year period from 1975 to 1995, the annual average amount of carbon stored in them will be much higher. This can be estimated using the general

Table 5.6 Net additional quantity of carbon stored by forestry plantations in Costa Rica during each of the years of the study (1), amount of carbon exiting the system every year (2), and net accumulated quantity through time (3)

Year	Hectares	(1)	(2)	(3)
1971	84	649	0	649
1972	168	1 298	0	1 947
1973	192	1 484	0	3 431
1974	213	1 646	0	5 077
1975	234	1 808	0	6 885
1976	239	1 847	0	8 732
1977	251	1 940	0	10 672
1978	285	2 203	0	12 874
1979	698	5 394	0	18 267
1980	1 505	11 630	0	29 897
1981	2 603	20 115	0	50 011
1982	3 960	30 600	0	80 611
1983	4 936	38 142	0	118 753
1984	6 222	48 079	0	166 832
1985	8 723	67 405	0	234 237
1986	12 898	99 666	0	333 904
1987	18 201	140 645	0	474 547
1988	23 036	178 005	0	652 553
1989	28 036	216 642	0	869 195
1990	41 833	314 212	9 044	1 183 406
1991	57 393	409 872	33 620	1 593 277
1992	73 351	543 716	23 087	2 136 994
1993	87 981	604 730	75 124	2 741 723
1994	102 609	771 246	21 642	3 512 969
1995	128 590	908 828	84 822	4 421 797
Average annual amount stored				746 769

procedure outlined above. However, the volumes to be harvested through time in the future would have to be estimated as well.

To estimate these volumes, it would be necessary to know the turn-to-harvest of the areas planted each year since 1971. The Ministerio de Recursos Naturales, Energía y Minas (1994a) reports that, in 1990, 41 per cent of the new areas established were planted with species with short turn-to-harvest (10–14 years, 12 on average), 19 per cent with species with medium turn-to-harvest (16–20 years, 18 on average) and 40 per cent with species with long turn-to-harvest (22–26 years, 24 on average).

During 1991, 1992 and 1993, those percentages changed to (50, 11, 39), (47, 9, 44) and (75, 3, 22), resulting in a four-year average of 53.25, 10.5 and 36.25 per cent, respectively. On the other hand, for all of the areas established from 1971 to 1995, the percentages are 40.7 and 53, respectively.

MINAE has not published detailed yearly data on the number of hectares of each species planted before 1990 and after 1993. Further, the estimation of the volumes to be harvested through time is greatly simplified if it can be assumed that the same relative mix of species, at least in regard to their turn-to-harvest, has been planted every year. Therefore a constant relative mix of 45, 10 and 45 per cent is assumed for the baseline estimations in this study.

It can be assumed that 45 per cent of the area planted in 1971 will, on the average, be harvested in 1983 and again in 1995, 2007 and so on; while 10 per cent will be harvested in 1989, 2007, 2025 and so on; and the other 45 per cent will be harvested in 1995, 2019, 2043 and so on. Repeating the former process for 1972 up to 1995, a hypothetical harvesting schedule can be obtained, ad infinitum, and the volumes to be harvested through time in the future estimated.

The calculations resulting in Figure 5.10 indicate that, if the existing 128 590 hectares of forestry plantations are maintained through managed natural regeneration or replanting in the same or different sites after harvesting, an annual average of 9 million metric tons of carbon would remain stored in

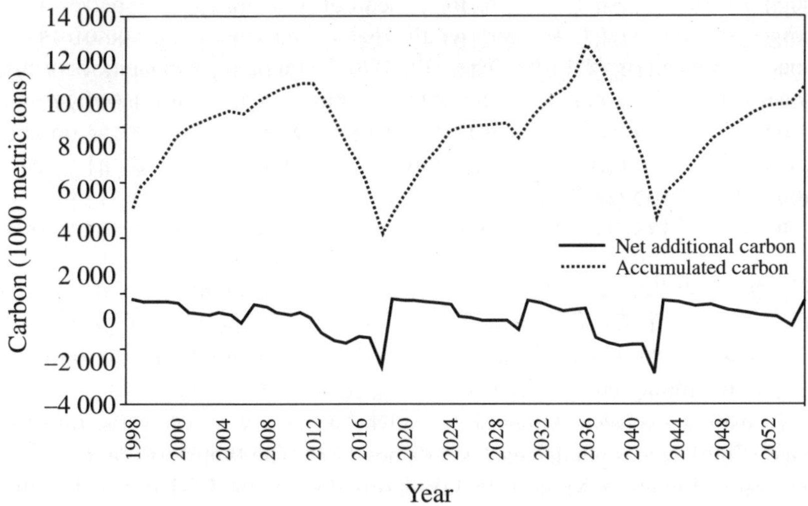

Figure 5.10 Carbon flows in Costa Rica's forestry plantation areas

the 20-year period from 1996 to 2015. This average is estimated at about 8.1 million tons for 2016–35, 8 million tons for 2036–55, and 8.3 million tons for 2056–75 (not shown in Figure 5.10). Therefore, over the next 80 years, a very stable average of 8.4 million tons of carbon per year would remain stored, just in the form of forestry plantations.

To provide an idea of the potential error that could be induced by the assumption about the relative mix of species, in regard to their turn-to-harvest (45, 10 and 45 per cent), it is estimated that an equal mix (33 per cent, 33 per cent, 33 per cent) would produce a relatively similar average of 8.7 million tons. If, on the other hand, all of the species planted had either a short, medium, or long turn-to-harvest, the corresponding averages would be of 5.6, 8.4 and 11.2 million tons per year. As expected, longer turns result in considerably higher average amounts of carbon remaining stored through time, amounts that are linearly proportional to the length of the turn (12:18:24 = 5.6:8.4:11.2).

Actually, there is a theoretical alternative to the former method of estimation, based on the premise that forestry plantations in Costa Rica sequester an average of 7.7 metric tons of carbon per hectare, per year. A hectare of a species with a short turn-to-harvest (12 years) thus would reach a maximum storage of 92.4 tons by the end of the twelfth year. On average, it would have retained 46.2 tons of carbon per year. Thus 128 000 hectares of this species in permanent rotations should maintain a yearly average of 5.9 million tons of carbon stored.

Similarly, species with medium turn-to-harvest (18 years) would reach a maximum storage of 138.6 tons for a mean of 69.3 tons, while species with long turn-to-harvest (24 years) would reach a maximum storage of 184.8 tons for a mean of 92.4 tons. Then 128 000 hectares in permanent rotations should maintain a yearly average of 8.85 and 11.8 million tons of carbon stored, respectively. The linear relationship between the length-of-turn and the yearly average quantity of carbon stored is illustrated, since 12:18:24= 46.2:69.3:92.4=5.9:8.85:11.8.

If 128 000 hectares of plantations were established during the same year with a relative mix of species with short, medium, and long turn-to-harvest of 45, 10 and 45 per cent, respectively, and maintained, ad infinitum, the yearly average amount of carbon permanently stored would be 128 000*0.45*46.2 +128 000*0.10*69.3+128 000*0.45*92.4=8 870 400, or 8.9 million tons.

The relatively small difference between the former figure and the 8.4 million tons estimate illustrated in Figure 5.10 is due to the fact that Costa Rica's forestry plantations were not all established during the same year and that such estimate was obtained considering a finite and relatively short time span. Also notice that the more involved calculations behind Figure 5.10 would yield a more precise estimate when the former was the case. Further, they must be used if plantation forestry areas are changing over time, which is by far the most common scenario.

Once the average amount of carbon that can be permanently stored in a given area is calculated, assuming that a relatively stable long-term productive land use pattern will be maintained, valuation is straightforward, conceptually. In practice, however, it is complicated by the fact that a liquid, stable, open market for carbon storage does not exist yet. However, a limited number of buyers and sellers have so far managed to come together and make transactions at a wide range of per-unit prices, as is expected in this kind of situation. In Costa Rica, for example, in the three projects already approved by the United States Initiative on Joint Implementation (USIJI), the CO_2 has been priced from US$10 to US$16 per metric ton, equivalent to US$36.6 to 58.6 per metric ton of carbon permanently stored, given the previously discussed conversion factor.

On the demand side, Carranza (1996) cites estimations of the cost of the damage caused by a metric ton of carbon released into the atmosphere of between US$10 and US$20 (or US$36.7 to US$73.2 per ton of carbon). Fankhauser and Tol (1995), cited by Carranza (1996), estimates that cost at US$20 per ton of CO_2 (US$73.2 per ton of carbon) and projects an increase to US$28 by the year 2000.

On the supply side, Winjum *et al.* (1993) reports the cost of storing one metric ton of carbon through plantation forestry at US$31 in Argentina, US$5 in Australia, US$10 in Brazil, US$11 in Canada, US$4 in Mexico and US$5 in the United States. Cline (1992), on the other hand, estimates these costs at US$5 per ton in tropical areas and US$20 in temperate areas of developed countries.

Emission control costs in industrialized countries are reported at between US$10 and US$20 per ton of carbon. The former figures provide a guideline for the general range of prices per ton of carbon stored that could be assumed for the purposes of this study. At the average of the prices already negotiated by the Costa Rican government with the USIJI, for example, the 8.4 million tons of carbon that can be permanently stored in the 128 000 hectares of plantation forestry already established in Costa Rica would have a one-time value of US$400 million. This is slightly lower than the estimate obtained when the mid-point of the range of cost-avoidance figures provided by Carranza (1996) (which includes Fankhauser and Tol, 1995) is used. On the supply side, the current average cost of storing one metric ton of carbon through plantation forestry of about US$11.5 (Winjum *et al.*, 1993; Cline, 1992) would imply a much lower value of US$97 million. If the current marginal cost of the alternative of emissions control is used instead, this estimate increases slightly.

CONCLUSIONS AND RECOMMENDATIONS

Humid tropical secondary forests have a considerable potential for carbon sequestration and storage. Most of the carbon sequestration takes place during the first 20 years of growth. The per hectare economic value of the carbon sink service provided exceeds the market value of the land in most tropical humid regions of the world. Therefore, if the market value reflects the opportunity cost of the land, forest conservation is an economically sensible alternative in these areas, especially considering that these forests are believed to render additional environmental services which are also economically valuable. Most of the carbon sink value relates to carbon sequestration, the preferred service for international joint implementation agreements.

If a global carbon market eventually develops, promoting secondary forest regeneration and conservation can yield substantial economic benefits for developing countries. Costa Rica has a relatively diversified non-agricultural economy and one of the highest per capita gross domestic products among developing countries. Even under these circumstances, receiving compensation for the carbon sink services of naturally regenerating secondary forest areas would represent a significant source of revenue for the economy.

In the case of primary forests, silvicultural treatments are found to significantly affect total biomass, and thus carbon stored. We found that the control and the upper canopy treatment store similar amounts of carbon during the first two production cycles, but the upper canopy begins to surpass the control during the third cycle. This does not cause a significant difference in the economic value of this environmental service in a long-term production scenario because of the detrimental effect of the discount rate on the value of money in the distant future. In contrast, the much lower long-term levels of carbon storage in the liberation treatment that remain after 60 years have an economic cost (that is, reduced payment) of around US$250/ha in relation to the control or upper canopy treatment.

The liberation treatment, however, shows a financial advantage in both the short and the long term, mostly due to its greater production of high-value wood during the first harvest cycle. Therefore, if forest owners get no payment for the carbon storage service or if the payment is at a fixed rate per hectare independent of the management scheme adopted and the resulting long-term storage levels, they will prefer the liberation treatment. Such a decision results in a net economic loss to society of between US$175 and US$300/ha, depending on the real discount rate assumed.

This translates to several hundred million dollars when considering the extension of natural forests areas even in a small country like Costa Rica. From the social standpoint, then, the control is the preferred alternative for the sustainable management of natural forests. It would be the treatment favoured by owners if they were paid correctly for the environmental service of carbon storage provided by their forests. None of the silvicultural treatments investigated are justified from the social viewpoint: liberation is handicapped by its negative effect on biomass, and upper canopy for its low impact on the commercial value of the harvests.

Costa Rica is currently paying owners US$360/ha for the sustainable management of natural forests. This is little, even when just compared to the economic value of the environmental service of carbon storage. According to this analysis, the Costa Rican payment is not sufficient to motivate long-term sustainable management of its primary forests. The advanced compensation mechanism adopted by Costa Rica, which provides 100 per cent of the payments for carbon storage during the first five years of a contract that has a 20-year duration, is also worrisome. The agreements have to be monitored and the necessary legal actions taken to ensure that the forest owners comply with the contracts. Otherwise, the most attractive alternative is to engage in a sustainable forest management contract, make a technically appropriate initial harvest, and sell or convert the land to another use five years later, once all contract payments have been received.

We find that, in both the short and the long term, the sustainable management of natural primary forests is only more profitable than selling or using the land for cattle or agriculture if owners are paid near the international rate of US$10/ton for the carbon storage service provided by their forests. The compensation mechanism would then become even more critical. If credibility can be established among forest owners with respect to the fulfilment of the payment agreements, it would be preferable to distribute them evenly throughout the 20-year forest production cycles, making the necessary annual adjustments to avoid the detrimental effects of inflation.

In the case of plantation forestry, the case study in this chapter demonstrates that the monitoring and valuation of at least one key environmental service, carbon storage, is feasible in a developing country, using currently available data and relatively simple methods. During the last 20 years, the rapidly growing forestry plantation areas in Costa Rica have kept an annual average of about 750 000 metric tons of carbon stored.

It is estimated that the 128 000 established up to 1995 could, on average, maintain 8.4 million metric tons of carbon in permanent storage. Given the terms of the current carbon bond initiative of the Costa Rican government, these could have a value of US$84 to US$168 million per 20-year period, or US$4.2 to US$8.4 million per year, equivalent to US$32.5 to US$65 per ha/year.

If the current trend of growing plantation forestry areas continues in Costa Rica, the former figures would be much higher. They could be estimated using the procedure advanced in this study. In addition, the previous calculations implicitly assume that, once the plantations are harvested, the carbon stored is immediately released into the atmosphere in the form of CO_2. It is important to estimate CO_2 'release curves' that predict the percentage of this gas, or its carbon equivalent, that will be annually discharged back into the atmosphere after harvest. These would allow for an upward revision of the former estimates that could be substantial.

Although an evaluation of Costa Rica's plantation forestry incentive programmes is not one of the objectives of this study, the results raise an interesting point about them. It is shown that the average amount of carbon that can be permanently stored in a hectare of plantation forestry is directly and linearly proportional to the length of the turn-to-harvest of the species planted. Therefore, if one of the objectives of or justifications for an incentive programme is to reduce net CO_2 emissions, this should be taken into account and the value of the incentives ought to be higher for species with longer turn-to-harvest. This is not the case in Costa Rica.

Overall, the analyses indicate that, if carbon sequestration is one of the main objectives of a system of forest incentives, promoting secondary forest regeneration should be a priority over primary forest conservation. This is not the case in Costa Rica and most other developing countries' incentive programmes. The fact that secondary forest ecosystems in their earlier stages of development are very vulnerable to land use change provides an additional argument for higher incentives.

A comparative economic analysis of secondary forests and artificial forestry plantations as carbon sequestration alternatives is recommended. A complete panorama of the joint implementation alternatives for a developing country, like Costa Rica, would also include mitigation measures in the energy and transport sectors. The cost-effectiveness of carbon sequestration in secondary forests and plantations versus measures to reduce CO_2 emissions adopting less energy-intensive technologies and fossil fuel substitutes needs to be evaluated as well.

Finally, it is important to stress that direct cross-country compensation from industrialized to developing countries for their forests' carbon parking or storage services implies that the latter have an implicit right to emit all their stored carbon and should be compensated for not doing so. The argument for this is that industrialized countries account for most of the excess CO_2 already in the atmosphere, on a per capita or a per area basis. This proposition is not universally accepted. However, within-country incentives to forest conservation might be justifiable to reduce the country's overall CO_2 emissions. In the global carbon market envisioned by the United Nations Intergovernmental Panel on Climate Change (IPCC), this could result in an international compensation or avoiding having to make payments for excessive emissions.

Direct cross-country compensation from industrialized to developing countries for carbon sequestration services is a globally accepted mechanism. The compensating country can use the sequestered carbon to offset its own emissions and still comply with international conventions. For industrialized countries, this type of joint implementation agreement may be less expensive than reducing internal emissions at the margin.

NOTES

1. The material presented is primarily drawn from the research of the contributing author in cooperation with Brian Finnegan and Manuel Gomez, Rosalba Ortiz and Lissette Rodriguez from the Centro Agronómico Tropical de Investigación y Enseñanza (CATIE).
2. The growth rate is assumed constant across forest sites because data about the earlier stages of forest growth (first 25 years) are only available for the 'Tirimbina' site.

REFERENCES

Andersen, L.E. (1996), 'A cost–benefit analysis of deforestation in the Brazilian Amazon', working paper, Department of Economics, University of Aarhus, Denmark.

Andrasko, K. (1990), 'El recalentamiento del globo terráqueo y los bosques: estado actual de los conocimientos', *Unasylva*, 41(163), 3–11.

Barres, H. (1993), 'Carbon-fixing and timber production in tropical Klinki Pine forest plantations', technical report, The Klinki Pine Project.

Brown, S. (1996), 'Papel actual y potencial de los bosques en el debate mundial sobre el cambio climático', *Unasylva*, 47, 2–10.

Brown, S. and N. Adger (1993), 'Forests for international off-sets: economic and political issues of carbon sequestration', Centre for Social and Economic Research on the Global Environment (CSERGE), University of East Anglia and University College of London.

Brown, S. and A. Lugo (1984), 'Biomass of tropical forests: a new estimate based on forest volumes', *Science*, 223, 1290–93.

Brown, S., J.R. Gillespie and A. Lugo (1989), 'Biomass estimation methods for tropical forests with applications to forest inventory data', *Forest Science*, 35(4), 881–902.

Carranza, C. (1996), 'Valoración de los servicios ambientales de los bosques costarricenses', Sistema Nacional de Conservación (SINAC) and Centro Científico Tropical (CCT), San José, Costa Rica.

Cline, W.R. (1992), 'The economics of global warming', Institute for International Economics, Washington DC.

Corporación de Fomento Ganadero (CORFOGA) (1995), 'El subsector de ganadería de carne en Costa Rica', Discussion paper, CORFOGA, San José, Costa Rica.

Costa Rican Office for Joint Implementation (OCIC) (1995), 'Notas Técnicas', San José, Costa Rica.

Dixon, J.A., L.F. Sara, R.A. Carpenter and P.B. Sherman (1994), *Economic Analysis of Environmental Impacts*, London: Earthscan Publications.

Faeth, P., C. Cort and R. Livernash (1994), 'Evaluating the carbon sequestration benefits of forestry projects in developing countries', World Resources Institute/ U.S. Environmental Protection Agency, Washington, DC.

Fankhauser, S. and R. Tol (1995), 'Recent advancements in the economic assessment of climate change costs', Working Paper no. 31, Center for Social and Economic Research on the Global Environment (CSERGE) and Vrije University, Amsterdam.

Fearnside, P. and W. Malheiros (1996), 'Carbon uptake by secondary forests in the Brazilian Amazon', *Forest Ecology and Management*, 80, 35–46.

Finnegan, B., L. Guillen and D. Quiros (1993), 'Guía de las acciones en el área demostrativa de la finca Tirimbina, cantón Sarapiquí, Costa Rica', CATIE, Programa de Producción y Desarrollo Agropecuario Sostenido, Turrialba, Costa Rica.

Goldberg, A.S. (1964), *Econometric Theory*, New York: John Wiley & Sons.

Houghton, R., D. Skole and D. Lefkowitz (1991), 'Changes in landscape of Latin America between 1850 and 1985, 2: Net release of CO_2 to the atmosphere', *Forest Ecology and Management*, 38, 173–99.

Judge, G.G., W.E. Griffiths, R. Carter Hill, H. Lutkepohl and Tsoung-Chao Lee (1985), *The Theory and Practice of Econometrics*, New York: John Wiley & Sons.

Kaimowitz, D. (1995), 'Cattle ranching expansion and secondary forest in Central America.' Instituto Interamericano de Cooperacion Agricola (IICA), San José, Costa Rica.

Krinsky, I. and A.L. Robb (1986), 'On approximating the statistical properties of elasticities', *Review of Economics and Statistics*, 68, 715–19.

Ministerio de Ambiente y Energía (MINAE) (1996), 'Ley Forestal No.7575', Alcance No. 21 a la Gaceta No.72 del 16 abril de 1996, San José, Costa Rica.

Ministerio de Recursos Naturales, Energía y Minas (1994a), 'Informe Anual 1993. Programa de Desarrollo Forestal', FDF DGF-DECAFOR, San José, Costa Rica.

Ministerio de Recursos Naturales, Energía y Minas (1994b), *Boletín Estadístico Forestal*, no. 5, DGF, San José, Costa Rica.

Ministerio de Recursos Naturales, Energia y Minas (1995), 'Estadísticas Relevantes del Sector Forestal', DGF, San José, Costa Rica.

Nabuurs, G.J. and G.M. Mohren (1994), 'Modeling analysis of potential carbon sequestration in selected forest types', *Canadian Journal of Forestry Research*, 25(7), 1157–72.

Quiros, D. and B. Finnegan (1994), 'Manejo sustentable de un bosque natural tropical en Costa Rica: definición de un plan operacional y resultados de su aplicación', CATIE, Serie Técnica, Informe Técnico No. 225, Colección Silvicultura y Manejo de Bosque naturales No. 9.

Ramirez, O.A. (1997), 'Estimation and use of a multivariate parametric model for simulating heteroscedastic, correlated, non-normal random variables: the case of corn-belt, soyabeans and wheat yields', *American Journal of Agricultural Economics*, 79 (February), 191–205.

Ramirez, O.A. and M. Gómez (1999), 'Estimación y valoración económica del servicio ambiental de almacenamiento de carbono', *Revista Forestal Centroamericana*, 27, 17–22.

Ramirez, O.A., B. Finnegan, L. Rodriguez and R. Ortiz (2000), 'Evaluación económica del servicio ambiental de almacenamiento de carbono: el caso de vu bosque humédo tropical bajo diferentes estrategias de manejo', in J.A. Dixon, L.F. Scura, R.A. Carpenter and P.B. Sherman (eds), *Análisis Económico de Impactos Ambientales*, Turrialba, Costa Rica: CATIE.

Richards, F.J. (1959), 'A flexile growth function for empirical use', *Journal of Experimental Botany*, 10(29), 290–300.

Schroeder, P.E., R.K. Dixon and J.K. Winjum (1993), 'Ordenación forestal y agrosilvicultura para reducir el dióxido de carbono atmosférico', *Unasylva*, 44, 52–60.

Solórzano, R., R. de Camino, R. Woodward, J. Tossi, V. Watson, A. Vásquez, C. Villalobos and J. Jiménez (1991), 'La depreciación de los recursos naturales en Costa Rica y su relación con el sistema de cuentas nacionales', Centro Científico Tropical (CCT) and World Resources Institute (WRI), San José, Costa Rica.

Thoranisorn, S., P. Sahunalu and K. Yoda (1991), 'Litterfall and Productivity of Eucalyptus Camaldulensis in Thailand', *Journal of Tropical Ecology*, 7(2), 275–9.

Torres, G., R. Luján and M. Pineda (1995), 'Diagnóstico técnico del proceso de producción forestal en plantaciones de pequeña escala en Costa Rica', Instituto Tecnológico de Costa Rica, Cartago, Costa Rica.

Winjum, J.K., R.K. Dixon and P.E. Schroeder (1993), 'Forest management and carbon storage; an analysis of 12 key forest nations', *Water, Air and Soil Pollution*, 70, 239–57.

6. Agriculture versus forestry in northern Saskatchewan[1]

Mohammed Dore, Surendra N. Kulshreshtha and Mark Johnston

INTRODUCTION

Background

The province of Saskatchewan is a major grain and oilseed-producing region of Canada. Much of this production, however, is destined for export, to eastern Canada and Europe in the earlier part of the 20th century, and to Far East Asian countries in the more recent period. Increased demand for Canadian grains, and relative shortage of land in the southern belt of the grain-producing regions, led to the movement of agriculture northwards. Such expansion was also facilitated by several policies of the various levels of government. Major policies would include settlement policies, where almost free land was given to prospective settlers, and the subsidy for transporting grain and oilseeds to various export locations, known as the Crowsnest Pass Agreement Rates. Other polices, such as fuel rebates and quota systems, also provided encouragement to agricultural production in the northern belt of the grain-producing region.

Settlement of western Canada and, in particular, northern Saskatchewan, witnessed the opening up of some 4 million hectares of previously uncultivated lands, some of which were under forests and other natural vegetation. Much of this land was cleared in the early part of the 20th century in response to short-term high prices of grain in the international grain market. Much of the agricultural activity on the marginal lands was made profitable only under agricultural subsidies noted above. It is the contention of this study that such actions were not socially optimal, particularly in light of present-day environmental concerns. If land use patterns had been decided in light of true agricultural productivity plus present-day environmental concerns, these marginal lands would not have been diverted into agricultural pursuits. This is because social gain from keeping these lands under forests would have exceeded the private gain of using the land for production of low-yielding grain and forage.

Objectives of the Study

The objective of this chapter is to examine the rationale of using marginal lands for agricultural production in the boreal forest transition zone of Saskatchewan, by comparing social gains from alternative land use patterns. This is accomplished by incorporating long-term sustainability considerations, which include both economic and ecological considerations. The criteria developed here are then applied in evaluating the more socially desirable land use pattern for northern Saskatchewan.

This study should be construed as a case study in the context of global change. Selection of alternative land use patterns is specific to the study region, which further conditions the selection of methodology adopted for comparing desirability of a given land use.

Organization of the Chapter

The chapter is organized as follows. The second section describes the study region, northern Saskatchewan, in terms of ecological and economic features. Included here is a historical description and analysis of the economics of grain farming and the forestry sector in the region. In the third section, methodology adopted and selected results for the study are described, followed in the fourth section by a presentation on policy considerations that arise from the results presented here.

SALIENT FEATURES OF THE STUDY REGION

The study was conducted in a forest–agriculture fringe area of northern Saskatchewan. This region is shown in Figure 6.1. Selection of the region was guided by the familiarity of the study team with the region, in addition to the recognition of the features of the existing land base that could support alternative land uses. A unique characteristic of the selected region is that commercial forestry and commercial agriculture coexist within this belt. For this reason, the region will be referred to as the forest–agriculture fringe area.

The Ecology of the Boreal Plains Region of Saskatchewan

The Boreal Plain Ecozone extends from the Peace River area in British Columbia to the southeastern corner of Manitoba. Its sub-humid, moderately cold climate is typified by cold winters and moderately warm summers, and is strongly influenced by continental climatic conditions. The Boreal Plain is underlaid by shale deposits and has flat to gently rolling topography. Soils are derived from glacial moraine and lacustrine surface deposits. Productivity is limited by soils and climate, and this ecozone represents the northern limit of commercially viable agriculture (Wiken, 1986).

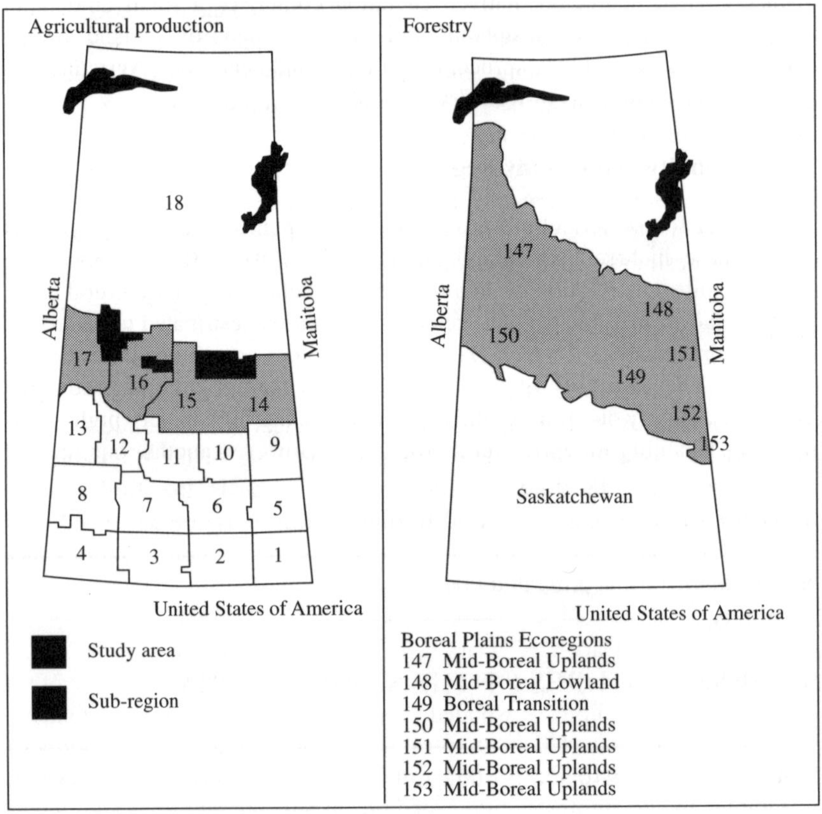

Figure 6.1 A map of northern Saskatchewan showing location of study region

In Saskatchewan, the Boreal Plain Ecozone has been divided into three ecoregions, as shown in Figure 6.1. From north to south, these are the Mid-Boreal Upland (MBU), the Mid-Boreal Lowland (MBL), and the Boreal Transition (BT) (SERM, 1995). As noted earlier, the focus in this chapter is on those lands in the MBU and MBL ecoregions on which agriculture is practised. While agriculture comprises 49 per cent of the land use in the BT ecoregion, it accounts for only 1.3 and 0.7 per cent in the MBU and MBL ecoregions, respectively (ibid.).

The MBU ecoregion is characterized by mixed deciduous–coniferous forest, dominated by early successional species such as aspen, jack pine and black spruce. Agriculture is limited to small areas of clay soils along rivers and streams where drainage has been improved.

In the MBL ecoregion, native vegetation comprises a small amount of mixed-wood forest on the uplands and extensive areas of lowland swamp with species such as black spruce, dwarf birch and peat moss. Agriculture is practised on small areas of well-drained uplands.

Soil Capability in the Study Region

The boreal forest–agriculture transition zone contains three major types of soils. These include Dark Gray and Dark Gray wooded soils, and Gray wooded soils. Distribution of area for these soil classes by agricultural capability is shown in Table 6.1. The region's land base is estimated to be around 4.1 million hectares, of which Dark Gray soil types constitute 60 per cent of the total. The capability of these various soils for agricultural production suggests that only one-third of the total land base has high to very high (Class 1 and 2) capability for agricultural production. However, another one-third of

Table 6.1 Distribution of soils in the study region

Soil capability for agriculture class	Area in '000s hectares			Total area in '000s (hectares)	% of total
	Dark gray soils	Dark gray wooded	Gray wooded		
1 and 2 (arable)	210	1 052	117	1 379	33.5
3 (marginal for crops)	136	400	681	1 217	29.6
4 (marginal)	17	67	111	195	4.7
5 – 6 (pastures)	108	496	667	1 271	30.9
7 and organic (unsuitable)	9	27	16	52	1.3
Total land area	480	2 042	1 592	4 114	100.0

Source: Shields *et al.* (1970).

the total land base is marginal for crop production, but may be suitable for pastures.

Economics of Alternative Land Uses

Growth of agricultural activity in the study region

Although the settlement of the districts of Saskatchewan and Assiniboia (later to be named as the province of Saskatchewan) started in the late 1800s, most of this occurred in the southeast part of the province, and around centres such as Moose Jaw, Swift Current and Saskatoon. As settlement of the province progressed, demand for land grew, and opening of the new areas became necessary. By the early 1930s, agriculture started pushing[2] into northern Saskatchewan, particularly in the forest–agriculture fringe area. In northern Saskatchewan, the area under agricultural pursuits increased steadily during the 1921–41 period (Figure 6.2).

Concurrently, the number of farms in the region also increased, from 24 488 in 1921 to 41 377 by 1931. A clearer picture of the opening of the northern Saskatchewan region can be seen in terms of the area that was brought under agricultural pursuits, as shown in Figure 6.2. In 1921, agricultural activity was confined to 2.76 million ha of land. In spite of the decline in the number of farms,[3] the corresponding area in 1966 increased to 5.6 million ha and virtually remained at that level until the present time.

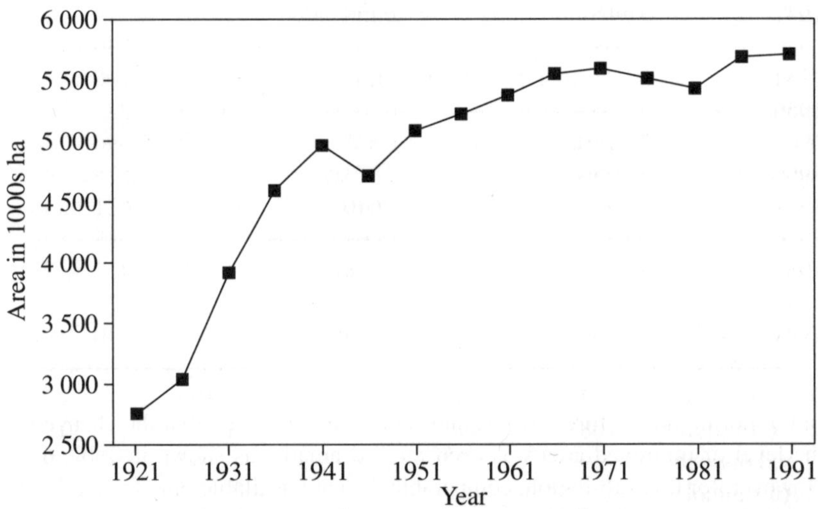

Figure 6.2 Area under agricultural pursuits, northern Saskatchewan, 1921–91

Another aspect of expansion of agriculture in northern Saskatchewan is conversion of previously unimproved areas on farms to cultivated area. This was done through clearing woodlands, converting native pasture to cropped area, and through draining areas under water (potholes, or wetlands). As shown in Table 6.2, total improved area, which includes area under crops and pastures, has increased steadily since 1921, from about one million ha then to 4.3 million ha by 1991. Although some of this increase has been through bringing additional areas under farming, a significant part of this is from converting unimproved lands on farms. Such lands reduced from 2.46 million ha in 1941 to 1.4 million ha in 1991. Based on the available data, it cannot be established whether the land being converted into cultivated area was marginal or of higher quality, although it is the contention of this study that it was the latter type of lands that was brought under agricultural production.

Table 6.2 Level of improved and unimproved area (in thousands hectares) under farms, northern Saskatchewan, 1921–91

Year	Improved area	Unimproved area	Total area
1921	1 071	1 691	2 762
1926	1 347	1 683	3 031
1931	1 888	2 027	3 919
1936	2 116	2 450	4 595
1941	2 516	2 458	4 979
1946	2 568	2 161	4 730
1951	3 035	2 065	5 100
1956	3 270	1 962	5 235
1961	3 518	1 875	5 396
1966	3 765	1 813	5 578
1971	3 949	1 670	5 621
1976	3 931	1 607	5 538
1981	4 129	1 344	5 473
1986	4 255	1 469	5 725
1991	4 306	1 434	5 740

In the northernmost forest-fringe sub-region, it was noted that the change in the level of improved area was even more dramatic, as shown in Table 6.3. However, for this sub-region, comparable data are available only for the 1921 and 1976 periods. In 1921, the total area under farms in the sub-region was 389 000 ha, which increased to 1.98 million ha by 1976. The unimproved area in 1921 constituted 71 per cent of the total, which decreased to almost

Table 6.3 Change in land use pattern in the northernmost fringe area, 1921–76

Land use category	1921	1976	Change in 1976 as % of 1921
Improved lands (ha)	112 890	1 394 739	1 235
Unimproved lands (ha)	276 920	587 218	212
Total area under agricultural pursuits (ha)	389 810	1 981 957	508
Unimproved area % of total area	71.0	29.6	—

30 per cent of the total by 1976. Thus much of the expansion in agricultural lands has been through clearing of the forest and wooded lands, which was carried out at a faster rate in the sub-region than that for the northern Saskatchewan study region as a whole.[4]

One of the major incentives for the opening up of northern Saskatchewan, and further expansion of agricultural pursuits, was high commodity prices received by farmers in the province. As shown in Figure 6.3, wheat prices during 1916–21 were relatively high. This encouraged further expansion of

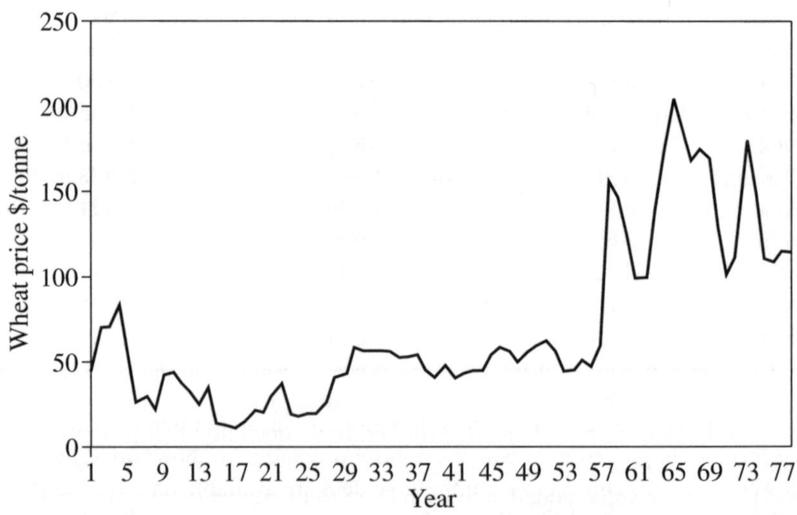

Figure 6.3 Trend in price of wheat received by farmers, Saskatchewan, 1901–77

agriculture in the province and, when the demand for land exceeded its availability in the settled agricultural areas in the southern part of the province, settlement and land claims moved northwards to lands in the study region.

Most of the improved lands in the study region are cropped lands (constituting 86 per cent of the total), although some smaller amounts of pasture lands and herds of cattle do exist. Furthermore, wheat is the 'king' of the crops, occupying roughly a fifth of the total cropped area. Since the most commonly followed rotation in the region is 'Fallow–Wheat', the wheat lands are roughly 40 per cent of the total cropped area. The next crops, in order of importance, are canola and barley. Genetic improvements in canola have made it possible for it to be grown in the northern fringe areas in the recent period. The other use of the improved lands in the region is for pasture. However, pastures constitute only 11.6 per cent of the total improved lands. Some limited amounts of livestock are also present on these farms.

Commercial forestry in northern Saskatchewan

Much of the northernmost part of the province of Saskatchewan is not suitable for agricultural pursuits. Much of this land is under forests, which occupy slightly over half of the province's total geographical area, or about 299 000 square km. However, the commercial forest cover is only 129 000 square km, of which only 77 000 square km (or 25.7 per cent of the total) is deemed productive forest.[5]

Although the Crown (provincial government) owns all forestlands, activities related to forests are undertaken by private companies, and are regulated by the provincial government through Forest Management License Agreements (FMLA). At the present time, the province has three FMLAs: (1) with NOR-SASK in the northwestern part of the province; (2) with Weyerhaeuser Canada Ltd, located in the area north of Prince Albert; and (3) with Macmillan-Blodel in the northeastern part of the province. Commercial forestry activity complements agroforestry activities in the province, which takes place on private woodlots. The province does not license many of the agroforestry activities, and thus details on them are scanty. However, the magnitude of such activities is not believed to be very large.

Commercial forests produce three types of timber products:[6] (1) wood pulp and paper products; (2) wood fabricated products, including lumber, plywood and waferboard; and (3) timber, fuelwood and other primary products.[7]

Much of the value-added activity related to the forest sector takes place in northern Saskatchewan. Although there are a limited number of sawmills in the southern part of the province, the number and capacity of such plants is very small. The areas around Prince Albert, Meadow Lake and Hudson Bay are the key centres for such activities. The province now has two pulp mills, one at Meadow Lake and the other at Prince Albert. The Prince Albert pulp mill is also integrated with a paper mill. Most of the products of the Saskatchewan forest sector are destined for exports outside the provincial boundaries, on account of a limited domestic market.

METHODOLOGY AND SELECTED RESULTS

Land Use and Global Climate Change

In the context of present-day concerns over environmental damage, land use planning should be consistent with the principles of environmental stewardship. The motivation of this study is the recognition of climate change (caused by emissions of carbon dioxide (CO_2) and methane into the atmosphere) as a global problem. In addition, it has been shown that the conversion of forest lands to agricultural lands is an important contributor[8] to increased atmospheric carbon dioxide, which is expected to lead to increases in global temperature and changes in precipitation patterns (Houghton *et al.*, 1987; Detwiler and Hall, 1988; Dale *et al.*, 1991).

In the past, land use decisions, in both the public and the private sectors, have been based simply on a criterion of private and short-run profitability, with no regard for its global ramifications or for changes that determine long-run sustainability of economic pursuits. Thus current land use decisions do not adequately take into account ecological characteristics of the land, and ensuring environmental benefits to society. This practice is inconsistent with the government's stated objective of adopting environmentally sustainable policies (Environment Canada, 1995). This inconsistency arises from the recognition that reforestation is an important tool in mitigating global warming because of its role in sequestering carbon (Sedjo, 1989; Dixon *et al.*, 1993; Freedman and Keith, 1995).[9]

Study Methodology

Development of sustainable land use requires the integration of ecological characteristics of forests and the resulting ecological services, together with factors that determine economic viability. One way of integrating ecological factors into land use decision making is to value the ecological services provided by forests.

In this study, evaluation of optimal land use pattern for the study region was conducted by comparing social benefits from each alternative land use. Two major economic activities in the study region – grain production and commercial forestry – were included. Although agroforestry is also practised in the region, given the small magnitude of such operations, and the scanty nature of information, its inclusion is left for future studies. The methodology involved a comparison of the flow value of services provided by forests (in terms of value added in forestry) to the flow value of alternative land uses, that is, grain production.[10] The flow value under a given land use was estimated through determination of net social benefits of each alternative land use. These net social benefits are expressed as time series over the sample period.

Let us define the 'total net social benefit' (TNSB) as a sum of two benefits: (1) private (commercial) net benefits and (2) social benefits from ecosystem services. In this study, ecosystem benefits are restricted to carbon sequestration.[11] The net benefits from a given land use are equivalent to the Marshallian concept of 'producer surplus', which reflects returns to all factors of production (land, labour and capital resources).

Let two flow service time series represent (a) flow service value of current use (use i), and (b) that in an alternative use, forests (use j). Thus, for a region k, alternative j is preferred over i if:

$$TNSB_{ik} \leq TNSB_{jk} \tag{6.1}$$

Comparison of two flow streams of benefits has conventionally been made in terms of a discounted capitalized stream of returns, the one with a higher net present value being chosen. This approach requires the selection of discount rates, use of which for environmental and long-term resource allocation is replete with problems. The innovative aspect of this study's methodology is that the two land uses were ranked using stochastic dominance. The use of stochastic dominance enables us to circumvent the well-known problems associated with the use of discount rates (see Sen, 1982; Parfit, 1984; Weitzman, 1994).

Stochastic dominance is based on the notion of stochastic efficiency, as formalized by Quirk and Saposnik (1962). The underlying assumption for the use of this technique is a monotonically increasing utility function, where the decision maker always prefers more to less. The stochastic efficiency of a distribution of net social benefits is assessed using cumulative probability function. If the cumulative probability function for the ith land use is located to the right of that for other land uses, land use i is preferred over the rest. This is called the first-order stochastic dominance criterion. If, however, the first-order stochastic dominance criterion does not lead to a clear choice or, in other words, the distribution for the ith option does not always lie to the right of jth distribution, second-order stochastic dominance is employed. For this criterion to provide a preferred choice, difference in the area under the distribution to the right of the other distribution should be larger than the corresponding area to the left.

Description of Variables and Sample Period

The analysis of the economics of grain farming in the study region was based on secondary data extending from 1961 to 1990. The region included the very northern forest–agriculture fringe area of the province of Saskatchewan, and included four Census divisions (CD): nos 14, 15, 16 and 17. Although some agricultural activity has begun to appear in CD18, the magnitude of such activities is rather limited and, for this reason, this CD was not included in the analysis. The selection of the sample period was determined primarily by the availability of data. For the period prior to 1921, although Saskatchewan as a region was identified in the Census, disaggregate comparable data were not collected. In the 1901 Census, the area designated as the province of Saskatchewan was shown as three districts: district of Athabasca (part), district of Saskatchewan, and district of Assiniboia. Thus details for the province were not available. In the 1906 Census, data were presented by township, whereas in the 1911 and 1916 Censuses data were reported by electoral districts, which are different from the Census divisions. For this reason, the analysis presented in this section is limited to the trends during the 1921–91 period.

To maintain comparability, data for commercial forestry were also collected for the same period. Net social benefits from agriculture on marginal lands were defined as the difference between gross value of production and its cost to society through resources used. However, the cost of production did not include a charge for labour and management or for interest cost on various types of farm investment. The gross value of production was simply a product of yield and price received by farmers. Two assumptions were made for this estimation: (1) that crop mix for the region has remained unchanged during the study period, and (2) that farmers on the marginal lands receive a price equal to the provincial average.[12] The yield on marginal lands was not available from any secondary data sources and, therefore, an expert opinion method was used. Contacts were made with the local extension officers. Their response suggested that yields on marginal lands might vary from year to year, and from land parcel to land parcel, within a range of 25 to 75 per cent of the average yield for the region. A mid-point of 50 per cent was selected for this analysis. All data were obtained from Saskatchewan Agriculture and Food (1991) and from Statistics Canada (1993).

For the forestry sector, net social benefits included two values. The first was the flow value of commercial production, which was estimated from secondary data sources. The forestry and allied products sector contained three major activities[13]: (1) logging of timber, (2) wood and wood products industries, and (3) pulp and paper industries. The first sub-sector produces an intermediate product, timber, which is used by the other two sub-sectors. Net social benefit from these three sub-sectors was estimated as the total value added by them during the sample period. A sum of these became the nominal value of net social benefits. Data for these sectors were obtained from Kulshreshtha *et al.* (1994).

The second net social benefit from forests is for ecological services, as noted earlier, limited to carbon sequestration benefits to the society. As explained in Dore and Johnston (this volume), this value is measured in terms of ability of forests to sequester carbon, and thereby mitigate global warming. In the Dore and Johnston study, value of carbon sequestration was estimated at a point where there is no externality; that is, when emissions of CO_2 are just equal to sequestration. This value of forests was combined with the flow value of forests for commercial services, to obtain a total social value of forests. All net social benefits were converted into constant dollar value using an implicit price index.

Selected Results

A comparison of the two net social benefits indicates that the land use under forests has a higher mean and a lower variance (Figure 6.4). The distribution

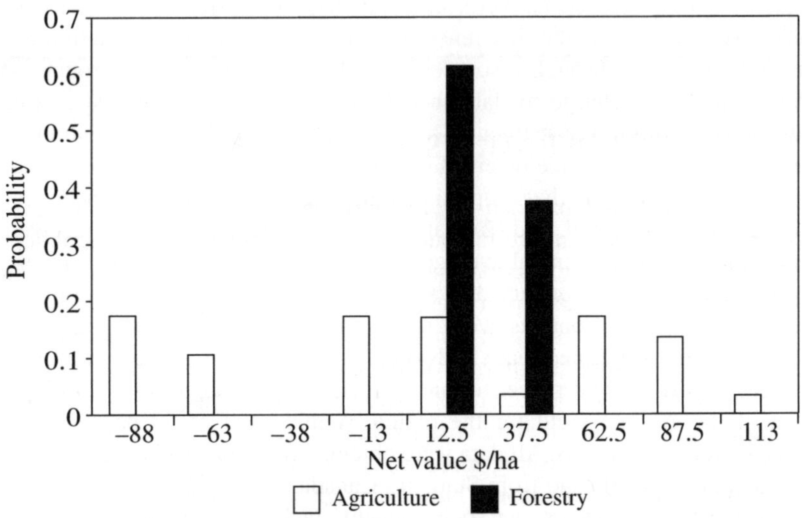

Figure 6.4 Net social benefits from grain production and forestry, northern Saskatchewan

for the forestland use indicated a positive social benefit for the entire sample period. The agricultural land use had a relatively higher variance. However, both distributions failed the normality test, and therefore it is clear from Figure 6.4 that neither distribution is normal. Therefore the procedure of second-order stochastic dominance was used to show which distribution should be socially preferred. Results for the cumulative frequency distribution, as shown in Table 6.4, show that the difference between the cumulative probability distribution of agriculture and forestry is positive. This suggests that land use under forests is preferred on the basis of this methodology.

Table 6.4 Results of the second-order stochastic dominance

Net social benefits	Probability		Cum. probability		Difference (agric – forestry)	Cum. difference
	Agric.	Forestry	Agric.	Forestry		
–99 to –75	0.173	0	0.172	0	0.172	0 .172
–74 to –50	0.103	0	0.275	0	0.275	0.447
–49 to –25	0	0	0.275	0	0.275	0.722
–24 to 0	0.173	0	0.448	0	0.448	1.17
1 to 24	0.172	0.62	0.62	0.621	0	1.169
25 to 49	0.03	0.38	0.654	1	–0.346	0.823
50 to 74	0.173	0	0.827	1	–0.213	0.61
75 to 99	0.139	0	0.966	1	–0.034	0.576
100 to 124	0.03	0	1	1	0	0.576
Sum	1	1				

POLICY CONSIDERATIONS

In areas of marginal productivity for agriculture, extensive subsidies are required to allow agriculture to continue. These include both financial support and large inputs of petrochemical-based fertilizers and pesticides. In examining the sustainability of these practices, other land uses become attractive alternatives. Conversion or rehabilitation of these areas by establishing forest cover represents an alternative that has both economic and ecological advantages. The exact magnitude of resulting carbon sequestration, although important, is a subject for another study. In this section, some of the ecological advantages in establishing forest cover on marginal agricultural land in the Boreal Transition Ecozone, either through natural forest succession or through plantations of commercially useful tree species, are explored.

Land use in the Boreal Transition ecoregion includes pasture, agriculture or forest. Ecological impacts of pasture include large inputs of nitrogen into streams and trampling of streamside areas, leading to bank failure and erosion. Agricultural impacts include soil erosion and loss of organic matter due to lack of permanent plant cover; soil and groundwater pollution due to high and continuous inputs of fertilizers and pesticides; and loss of soil organic matter from conventional tillage practices. Forest cover maintains soil stability, cycles carbon and nutrient elements and does not require large applications of chemicals.

Carbon in ecosystems is sequestered in soil organic matter and vegetation. Forests store large amounts of carbon because of the high levels of biomass in long-lived, large woody species, and because soils are high in organic matter and are infrequently disturbed. Boreal forests hold 30 to 80 tonnes C ha in vegetation and 200 to 500 tonnes C ha in soils (Dixon *et al.*, 1994). Johnston *et al.* (1996) analysed the effects of converting upland forest to agricultural land in the near-boreal forests of central Minnesota. They found that vegetative C declined from 50 tonnes C ha to 1 to 2 tonnes C ha and C stored in forest floor and the upper floor and the upper 25cm of mineral soil (the portion most affected by cultivation) declined from 40 to 25 tonnes C ha. Overall, C storage declined by 63 tonnes C ha, a decrease of 70 per cent. Schlesinger (1986) reviewed a number of studies of the effects of converting temperate forests to agriculture and found that soil C storage declined by an average of 30 per cent.

Van Kooten *et al.* (1992, p. 131) have estimated that marginal agricultural land area on the Canadian prairies is 4.41 million ha. They also estimated the current unimproved agricultural land area at 4.16 million ha. Both of these landmasses will be suitable for reforestation. Using the average rate of 63 tonnes of C ha, conversion of these lands to forests would lead to 277.83 million tonnes of C sequestered from the marginal lands, and another 262.08 million tonnes from the unimproved lands, for a total of 539.91 million tonnes of carbon.

Given the commitments made by the Canadian government to the Framework Convention on Climate Change in Rio de Janeiro in 1992, and at the Kyoto Protocol in 1997, to reduce emissions of greenhouse gases (of which carbon dioxide is the major gas), the conversion of agriculture to forest represents a large gain in C sequestration and, through that, reduction of carbon dioxide. Forests can be established on these lands either by allowing natural successional processes to occur, or by establishing plantations of fast-growing, commercially valuable species. Natural succession is attractive from the biodiversity point of view, providing maximum habitat for a range of plants and animals. Assuming that the previous agricultural activities did not severely reduce site productivity, natural succession will also result in a forest similar to the one that existed before clearing. Plantations will provide greater C sequestration, due to the use of fast-growing species and to the fact that trees are planted at optimum density. Aspen, in particular, fixes C at a high rate. In addition, plantation species are chosen for their commercial value, so the economic returns from plantations are generally greater than those from naturally regenerated stands. This is especially true of abandoned agricultural lands, as establishment costs are low and site productivity is often high. For example, in northern Saskatchewan, species used by the pulp and sawmilling industries are shown in Table 6.5. These data suggest that some combination of aspen, white spruce and jack pine would provide the best mixture, depending on whether C sequestration or economic return is of greater value. However, biodiversity is generally less in plantations because only a single species is planted and because the higher density of trees casts heavy shade and prevents the establishment of other plant species.

Table 6.5 Composition of timber harvested for pulp and lumber production, northern Saskatchewan

Pulp		Sawmilling	
Softwoods (%)			
White spruce	23	White spruce	80
Black spruce	17	Jack pine	12
Jack pine	16	Balsam fir	5
Balsam fir	2	Black spruce	3
Hardwoods (%)			
Aspen	34		
Balsam poplar	7		
White birch	1		

Reforestation of marginal lands is not only desirable for the mitigation of global warming, but also from another economic consideration. Reforested land can be purchased by energy-producing industries to earn carbon credits. An alternative to such a reforestation in Canada would be the purchase of existing forest stands outside Canada which have economic repercussions for the local area, and for communities living there.

In summary, this proposal is a local solution to a local problem of the optimal land use policy for marginal lands. It is also a component of a solution to the global warming problem. In addition, the switch in land use from agriculture to forests would lead to two other benefits: (1) there would be a reduction in CO_2 emissions through a reduction in the burning of fossil fuels; (2) it would enable major fossil fuel-burning industries in Canada to earn 'carbon credits' by financing such reforestation. This would further lead to Canada fulfilling its international treaty commitments under the United Nations Convention on Climate Change, 1992, as amended in Berlin, 1995. A further benefit is that foreign exchange would be saved if Canadian corporations could earn the carbon credits in Canada rather than in a foreign country. The reforestation could generate employment that might be particularly welcomed by the aboriginal peoples of northern Saskatchewan.

NOTES

1. This chapter is a revised version of a paper presented at the Annual Meeting of the Canadian Resource and Environmental Economics Study Group, McGill University, 5–6 October 1996. The authors are grateful to Professor Chris Green of McGill University, and to Professor Andy Muller of McMaster University, for many valuable comments on an earlier draft of this chapter.
2. During the 1911–31 period, agricultural activity also moved into southwest Saskatchewan.
3. The post-world war II period witnessed a dramatic decline in the number of farms, for several reasons, including rapid technological change biased towards labour-saving practices, change in the definition of census farms, and change in the aspirations of farmers for a better life.
4. This is supported by the fact that the increase in the improved area (that was brought under agricultural pursuits) during 1921–76 is estimated to be 12.3-fold for the northern fringe area, as against only 3.7-fold for northern Saskatchewan as a whole.
5. The distinction between productive and non-productive commercial forestland is based on accessibility. Under the present institutional arrangements, and available infrastructure, some forests are not deemed to be profit-generating, and are called 'unproductive'.
6. In addition to the timber products, forests also produce non-timber products. Such products include recreation and ecotourism, grazing of livestock, trapping, hunting and speciality forest products.
7. One should note that timber included under the third category of products is used as an input for the other two types of products.
8. Agriculture contributes to atmospheric carbon dioxide through three major sources: (1) through deforestation, which directly causes the release of CO_2 into the atmosphere; (2) tillage practices followed in pursuit of agricultural production release soil CO_2 into the atmosphere through depletion of soil organic matter; and (3) through use of mechanical technology, which burns fossil fuels, thereby adding further to the emissions of CO_2 to the atmosphere.
9. For example, the establishmet of poplar plantations on abandoned agricultural land has been shown to sequester 1.5–2.5 tonnes C ha/yr (Nabuurs and Mohren, 1994; Schlamadinger and Marland, 1996).
10. Although some livestock production does take place in the region, it was assumed that the scale of such activity is not significant, and therefore would not affect the conclusions of this study.
11. In addition to carbon sequestration, forests also provide other ecological services, such as water quality improvement, and maintenance of natural habitats such as for wildlife and other natural species. All these services affect human welfare and are therefore important. However, these services are not estimated in this study.
12. This assumption may be somewhat optimistic, since the study region is situated in the northern part of the province, and the cost of transportation to export destinations may be somewhat higher than that paid by an average garmer in the province.
13. In addition to these three sub-sectors, paper and allied industries are also related to the

Saskatchewan forest products. However, the contribution of forest products to the total value of final output for these industries is rather small. For this reason, these industries were not included.

REFERENCES

Dale, V.H., R.H. Houghton and C.A.S. Hall (1991), 'Estimating the effects of land use change on global atmospheric CO_2 concentrations', *Canadian Journal of Forest Research*, 21, 87–90.

Detwiler, R.P. and C.A.S. Hall (1988), 'Tropical forests and the global carbon cycle', *Science*, 239, 42–7.

Dixon, R.K., K.J. Andrasko, F.G. Sussman, M.A. Lavinson, M.C. Trexler and T.S. Vinson (1993), 'Forest sector carbon offset projects: near-term opportunities to mitigate greenhouse gas emissions', *Water, Air and Soil Pollution*, 70, 561–77.

Dixon, R.K., S. Brown, R.A. Houghton, A.M. Solomon, M.C. Trexler and J. Wisniewski (1994), 'Carbon pools and flux of global forest ecosystems', *Science*, 263, 185–90.

Environment Canada (1995), *The 1995 Report to the United Nations Commission on Sustainable Development*, Ottawa, Ontario: Environment Canada,

Freedman, B. and T. Keith (1995), *Planting trees for carbon credit*, report prepared for Tree Canada Foundation, Ottawa, Ontario.

Houghton, R.A., R.D. Boone, J.R. Fruci, J.E. Hobbie, J.M. Melillo, C.A. Palm, B.J. Peterson, G.R. Shaver, G.M. Woodwell, B. Moore, D.L. Skole and N. Myers (1987), 'The flux of carbon from terrestrial ecosystems to the atmosphere in 1980 due to changes in land use: geographic distribution of the global flux', *Tellus*, 39B, 122–39.

Johnston, M.H., P.S. Homann, J.K. Engstrom and D.F. Grigal (1996), 'Changes in ecosystem carbon storage over 40 years on an old-field/forest landscape in east-central Minnesota', *Forest Ecology and Management*, 83, 17–26.

Kulshreshtha, S.N., J. Siemens, W. Doell and H. Walker (1994), *Economic Perspective on the Prince Albert Model Forest Region of Saskatchewan*, Prince Albert: The Prince Albert Model Forest Association.

Nabuurs, G.J. and G.M.J. Mohren (1994), 'Modeling analysis of potential carbon sequestration in selected forest types', *Canadian Journal of Forest Research*, 25(7), 1157–72.

Parfit, D. (1984), *Reasons and Persons*, Oxford: Oxford University Press.

Quirk, J.P. and R. Saposnik (1962), 'Admissibility and measurable utility functions', *Review of Economic Studies*, 29, 140–46.

Saskatchewan Agriculture and Food (1991), *Agricultural Statistics, 1990*, Regina: Province of Saskatchewan.

Schlamadinger, B. and G. Marland (1996), 'The role of forest and bioenergy strategies in the global carbon cycle', *Biomass and Bioenergy* 11(6), 499–500.

Schlesinger, W.H. (1986), 'Changes in soil carbon storage and associated properties with disturbance and recovery', in J.R. Trabalka and D.E. Reichle (eds), *The Changing Carbon Cycle: A Global Analysis*, New York: Springer Verlag.

Sedjo, R.A. (1989), 'Forests to offset the greenhouse effect', *Journal of Forestry*, 87, 12–16.

Sen, Amartya (1982), 'Approaches to the choice of discount rates for social benefit–cost analysis', in: R. Lind (ed.), *Discounting for Time and Risk in Energy Policy*,

Washington, DC: Resources for the Future; reprinted in Amartya Sen (1984), *Resources, Values and Development*, Oxford: Basil Blackwell.

SERM (1995), 'The boreal plain ecozone: a forest community', *Saskatchewan's State of Environment Report*, Saskatchewan Environment and Resource Management, Regina.

Shields, J.A., H. Rostad and J.S. Clayton (1970), *Inventory of Saskatchewan Soils and Their Capability for Agriculture Use*, Saskatoon: Saskatchewan Institute of Pedology, University of Saskatchewan.

Statistics Canada (1993), *Census of Agriculture*, Ottawa.

Van Kooten, G.C., L.M. Arthur and W.R. Wilson (1992), 'Potential to Sequester Carbon in Canadian Forests: Some Economic Considerations', *Canadian Public Policy*, XVIII(2), 127–38.

Weitzman, M. (1994), 'On the "Environmental" Rate of Discount', *Journal of Environmental Economics and Management*, 26, 200–209.

Wiken, E.B. (compiler) (1986), *Ecozones of Canada*, Ecological Land Classification Series No. 19, Ottawa: Environment Canada, Lands Directorate.

7. Economic valuation of mangrove ecosystems and sub-tropical forests in Central America

Tania Ammour, Néstor Windevoxhel and Gustavo Sención

INTRODUCTION

The loss of tropical biodiversity through land use changes in developing countries is a problem which continues to make headlines throughout the world. Although studies and research have been carried out and applied, to some degree, in developed countries, the economic planning of developing countries does not yet incorporate or evaluate the impacts of management strategies beyond the goods produced. This has to do with, among other reasons, the unavailability of reliable information for tropical ecosystems, as well as the need to adjust methodologies to specific ecological and economic conditions found in developing countries.

The need to translate ecological information into economic terms is immediate, inasmuch as decision makers in these countries have begun to recognize the importance of environmental services of their natural resource stocks, as reflected in Costa Rica's recent agreement with specific environmental services such as tourism, carbon fixation and so on. While the evaluation of the ecological function of carbon fixation and other individual services is a step forward, it is necessary to value, not just the environmental services, but resource management practices and their effect on these services. However, there are few studies on partial or total ecosystem valuation in Latin America (Barton, 1995; Furst *et al.*, 1996; Pearce *et al.*, 1993). Considering the critical poverty levels in developing countries, the primary concern is the creation of value through the production of goods respecting the basic ecological processes of ecosystems. In practice, it is not possible to separate conservation from development objectives.

In this sense, the purpose of the evaluation of ecosystem sustainability does not only have academic ends but should be used as a tool for decision makers within developing countries as well as internationally. In particular, valuation studies provide technical and economic criteria for those responsible for the orientation of development planning as well as natural resource management research. In the latter case, there is little useful information on ecological processes which can be applied in valuation studies. Consequently, there is a need to adapt and generate methods which quantify the effects of management practices on ecosystem conservation and apply them in specific situations.

The Tropical Agricultural Research and Higher Education Center (CATIE) through the project 'Conservation for sustainable development in Central America', known as Olafo, is designing, with local communities, sustainable natural resource management activities in natural ecosystems. Since 1989, field activities have been carried out in five Central American countries (Guatemala, Honduras, Nicaragua, Costa Rica and Panama) and include research, training and technical assistance. The design and validation and promotion of sustainable management of timber and non-timber forest products focus on different hierarchical levels of complexity: species, components (such as forests), family production systems and area of influence at the community or landscape level. Proposed management practices should be measured at the landscape level to evaluate their sustainability as well as their contribution toward reconciling conservation and development.

In 1991, the Olafo project, together with the World Conservation Union (IUCN), the International Institute for Environment and Development (IIED) and the University of Maryland, has designed guidelines for the economic evaluation of tropical wetlands (Barbier *et al.*, 1993). These guidelines were applied to different ecosystems in Central America, adjusting for local social and economic conditions. In particular, the methodology has been used in four cases in two countries: mangroves in Nicaragua, broadleaf forest and agroecosystems in the subtropical forest in Guatemala, and a flood plain wetland in Guatemala (Petexbatun).

This chapter includes the methodological and specific results of two of the case studies which focused on forest valuation: mangroves of the Pacific coast in Nicaragua and broadleaf forest in the northern part of Guatemala. Despite the fact that the two ecosystems evaluated are different in terms of their ecological, economic and social characteristics, they have similar problems:

- they are subject to extractive pressure from local and external populations. In practice, the natural resources are open-access goods;
- productivity of the most commonly used timber resources is relatively low: in Nicaragua, *Rhizophora mangle* does not naturally resprout; in Guatemala, *Cedrela odorata* and *Swietenia macrophylla* have relatively slow growth rates;
- extraction practices are selective and limited to a few species with actual commercial value.

In the case studies presented, the value of ecosystems is not inherent in itself but, rather, depends on the harvesting capacity, which varies according to management practices. In this sense, the evaluation of different management scenarios is useful to orient decisions.

The hypotheses in these two cases include the following: (1) the sustainable management of these forests provides a higher value than traditional extractive practices, (2) the sustainable management of goods implies that ecological functions create more value, and (3) ecological functions provide more value than goods. The objectives were to estimate the partial economic value of each ecosystem based on field data generated in productive ecosystems, and to develop a methodology.

THE STUDY AREAS

The Mangroves of the Pacific Coast of Nicaragua

The mangroves of the Pacific coast of Nicaragua cover 393km^2, of which 22 per cent (87km^2) corresponds to a mangrove area called Héroes y Mártires de Veracruz, located between the Isla Maderas Negras and the community of Salinas Grandes in the region of León–Chinandega (see Figure 7.1). The mangroves on the Pacific coast of Nicaragua are found in the dry tropical forest life zone where the dry season lasts at least five months. The average

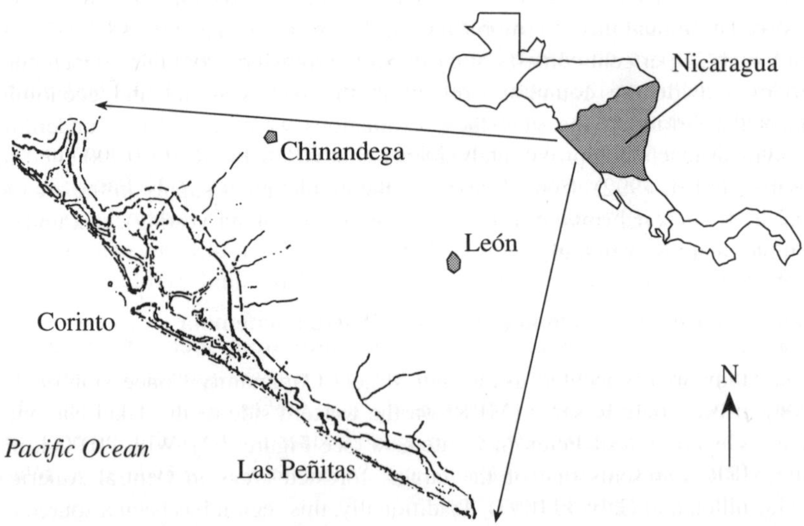

Figure 7.1 Mangroves of Pacific coast, Nicaragua

annual temperature is 26°C (Prado and Silva, 1992, cited in Windevoxhel, 1992).

The mangroves structure is dominated by fringe mangroves (Lugo and Snedaker, 1994). Periphery vegetation along estuaries parallel to the seashore is dominated by *Rhizophora spp.* with some *Laguncularia racemosa.* Behind these are fringes dominated by *Avicennia spp.* which are characteristic of border mangroves in low precipitation areas. The forest is characterized by poor structural development (low tree height ranging from 7 to 11m). Mangrove productivity is limited, among other factors, by low nutrient absorption, high salinity, low water recharge flow (rivers are intermittent) and high sediment anaerobiosis (Windevoxhel, 1994).

Topography is flat and predominant soils are clayey and easily flooded. In the mangroves, the courses of the rivers create networks of dendrytic canals in swampy areas and some rivers reach the sea through estuaries and canals (Rojas, 1987, cited in Windevoxhel, 1992). Some sandy islands bordered by mangroves and estuaries form beaches, often used for recreation. One of the most popular zones for tourism is Poneloya-Las Peñitas, 19km away from the main city of the Pacific region, León (Windevoxhel, 1992).

Of the total population located in the area (33 300 inhabitants), 30 per cent are directly dependent on mangrove resources for fuelwood and seafood, according to the Olafo project's preliminary evaluations. Approximately 90 per cent of the total population extracts fuelwood from the mangroves as an energy source. The extractive activities (fishing, fuelwood, shellfish, bark for tannins) are activities complementary to salaried agricultural and service work. The annual income (monetary and non-monetary) averages US$1900 – of which 21 per cent consists of cash income transfers from family members living outside the country – reflecting the zone's social and economic marginalization. At the time these evaluations were carried out, underemployment reached approximately 24 per cent (IICA/FLACSO, 1990 cited in Windevoxhel, 1992). Considering that the employment supply in the area is basically seasonal (mainly for sugar cane and cotton harvesting), real underemployment is even higher.

The San Miguel Subtropical Forest in Petén, Guatemala

The study area is located in the San Miguel Community Concession of the Mayan Biosphere Reserve (MBR) on the western side of the Tikal National Park, Department of Petén in Guatemala (see Figure 7.2). With 20 000km^2, the MBR represents one of the largest forested areas in Central America (Marmillod and Galvez, 1997). Traditionally, this region has been a source of valuable timber species and non-timber products such as latex, allspice and ornamental palms.

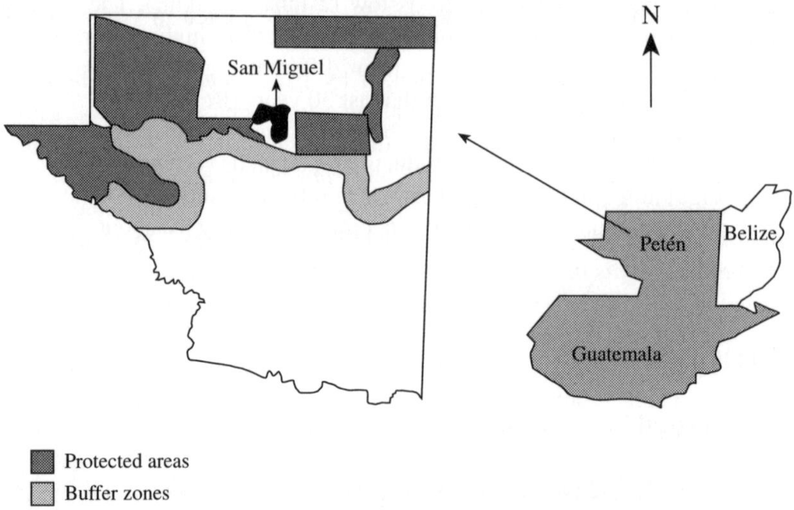

Figure 7.2 Community concession of San Miguel: Petén, Guatemala

In 1994, the community of San Miguel was granted usufructuary rights to 7000 ha of forest and agricultural land by the National Council of Protected Areas (CONAP), the government institution responsible for the administration of the Biosphere Reserve management. While the forest is managed collectively, agricultural lands are cultivated on an individual/family basis. According to the contract signed between the Guatemalan government and the community for a 40-year period, the latter is responsible for the sustainable management of these natural resources.

The San Miguel Community Concession (SMCC) is found in a subtropical forest life zone where the dry season lasts at least five months. The average annual temperature is 23°C and annual average precipitation is 1530 mm (Marmillod and Galvez, 1997) The SMCC is located in a calcareous formation. The area topography is characterized by slopes which alternate from medium stony rolling plateau to very steep stony hills. Drainage is adequate at the top and slopes of hills (Rendzinas) but poor in the lower valleys (Pellic Vertisols) (Collinet, 1997). The forest, while made up of more than 100 tree species, is dominated by 10 species which represent 50 per cent total forest biomass[1] (Marmillod and Galvez 1997) . From a silvicultural point of view, forest homogeneity is an advantage because it simplifies management; however, from all these species, only one (*Manilkara zapota*) generates products with a current commercial value.

The area is representative of tropical agricultural frontier areas and has the following characteristics: 6400ha of forest cover represent 91 per cent of the total concession, population density is low (2 inhabitant/km^2). The level of illiteracy and of social and organizational conflicts is high as a result of relatively recent migratory processes from different regions of Guatemala, Mexico, Belize and El Salvador over the last 30 years (Reyes, 1996) .

Traditional production systems are based on slash and burn agriculture, selective extraction of non-timber products (ornamentals such as *Chamadorea spp*.–xate), hunting and home gardens for family consumption. The average annual monetary and non-monetary income per family is about US$700 (ibid.).

METHODOLOGY

Methodological Framework

The methods used in the present study followed those developed by Barbier *et al.* (1993) together with CATIE and IUCN to value tropical wetlands. According to these guidelines, ecosystems can be valued according to three alternative ways: impact valuation, partial valuation or total valuation (see Figure 7.3). The selection of a particular type of valuation depends upon the problem definition which defines the scope of the study (ibid.):

- a specific environmental impact (for example, sea oil spilling) can be the cause of ecosystem damage. In this case, only changes produced by this action should be valued, and the valuation corresponds only to an impact ecosystem evaluation;
- different management options are identified and some are selected. Ecosystem evaluation is partial because it considers only some of the resources and environmental services. The decision of applying one management alternative should be compared with the opportunity costs of proposed options;
- complete change in the ecosystem use/management is considered (that is, the creation of a protected area), or the objective is to measure, for example, the economic value of the ecosystem for the society as a whole (that is, natural resources accountancy). In these cases, the total net benefits of the ecosystem should be valued, including all the net production and environmental values.

It is recognized that ecosystems have direct use (goods), indirect uses (services and ecological functions) and non-use value (attributes). In the present chapter, 'ecological functions' and 'environmental services' are considered to be synonymous. Both concepts pertain to the generic definition of

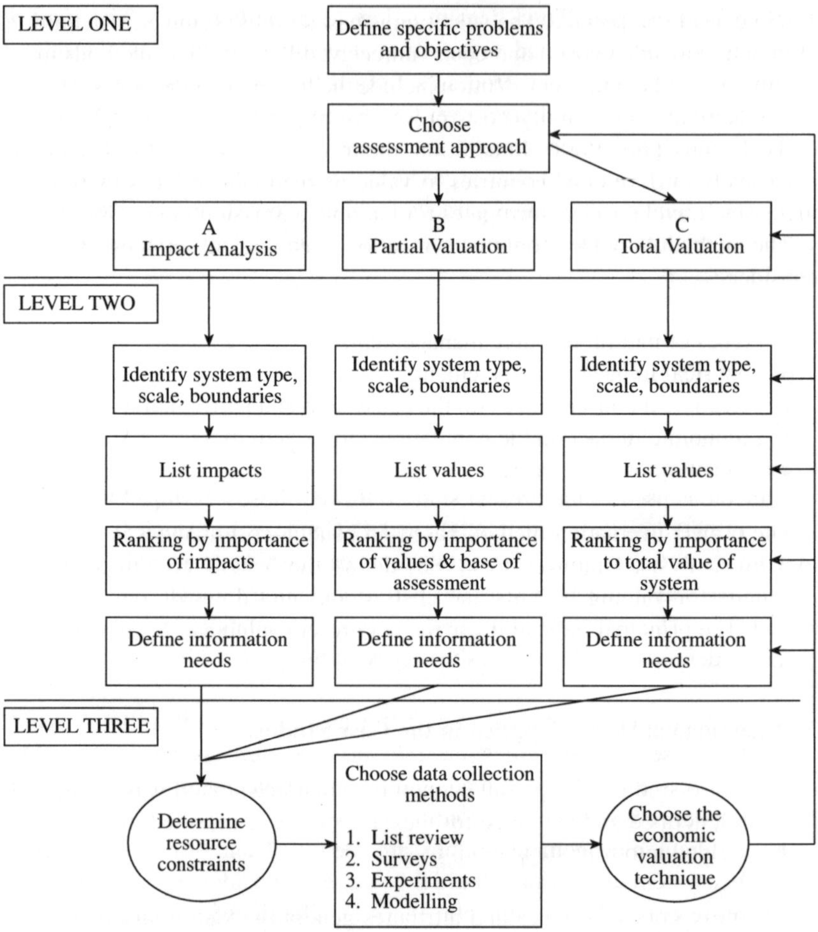

Figure 7.3 Wetland evaluation decision flow chart

'services'. Economic and information services such as recreation, tourism or transportation are considered separately.

The economic evaluation of net benefits produced by ecosystems goods is calculated using standard economic valuation, allowing for the quantification of direct incomes and extraction/production costs.

In order to value indirect uses, it is necessary to identify and quantify – within and outside the studied ecosystem – the direct damages or benefits associated with the hypothetical loss of the selected ecological functions or services. Identifying the ecological impact of direct ecosystems use allows for the measurement of the indirect effects of damages/improvements caused by the considered service. The valuation is based on the symmetric definition of benefits (avoided costs) and costs (unreceived benefit). The methods most commonly used in indirect valuation include the following approaches: change in productivity, opportunity cost, preventive expenditures and replacement costs (Dixon *et al.*, 1988). In addition, the travel cost method has been used extensively in developed countries to value recreational services by treating increasing travel costs as surrogates for variable admission prices (ibid.).

The methodology for economic evaluation of ecological services explicitly considers:

- types of natural resource management,
- physical impact of management,
- ecological consequences on the ecosystem and surrounding area,
- economic impact on the population and its activities,
- valuation of each service.

The project evaluation method based on benefit–cost analysis is used as a tool for analysis. Dixon *et al.* (1988) define the benefit–cost method as a technique for valuing benefits and costs stocks and flows during a defined period. The principal indicators are the net present value and the benefit–cost ratio.

Methodological Design Applied in the Case Studies

In both case studies, the partial economic valuation method was applied. A flow chart defined and validated for the case studies is reported in Figure 7.4 and is divided into four steps.

1. Identify goods, services and attributes generally existing in the ecosystem. Table 7.1 describes them for tropical broadleaf forests and mangroves ecosystems.
2. Rank goods, services and attributes according to their actual and potential value in the area, and to both their economic and ecological importance in the ecosystem.
3. Select goods, services and attributes to be valued and define valuation methods according to data availability and quality.
4. Value goods, services, attributes and ecosystem. The stream of each good and service net benefit value over time allow for approximating the partial value of the ecosystem according to the considered management scenario.

Limitations to the methodology include the possibility of double accounting of the values when adding individual goods and services, and the

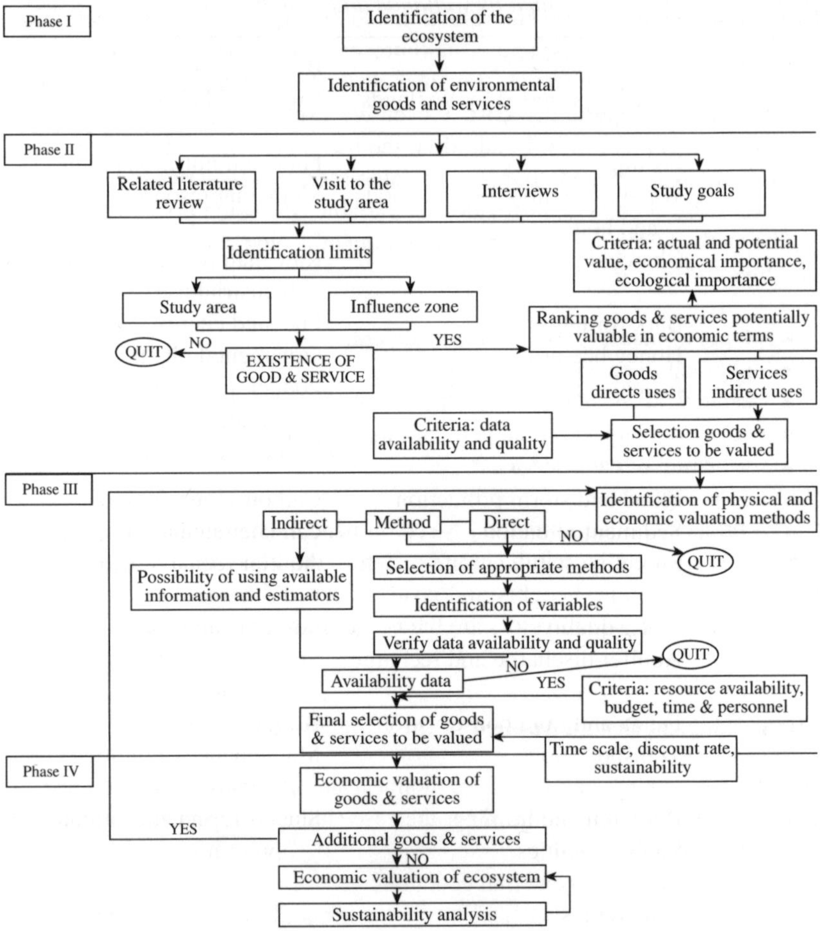

Figure 7.4 Methodology for economic valuation of tropical ecosystem

simplification of ecological processes over time and geographically. On the other hand, the methodology allows for the design of independent natural resource management scenarios and, provided with the available information, can be used to elaborate sensitivity analysis. This characteristic is particularly suitable in developing countries where reliable information is not always available.

In the case of Nicaragua, projections were made to determine the net present value for both traditional and improved (sustainable) management scenarios over time. Sensitivity analysis was applied to crucial variables such as labour costs and prices. In the Guatemala case, projections were not made

Table 7.1 Benefits provided by wetlands and broadleaf forests

	Benefits provided by Wetlands	Benefits provided by Broadleaf Forest
Goods	Fishing	Food and nutritious drinks
	Wood extraction for fuelwood and housing	Medicinal resources
	Charcoal production	Raw material for housing and industrial use
	Bark for tannin	Fuel and energy
	Honey bee production	Ornamental plants
	Fauna and flora harvesting	Fodder and fertilizer
		Raw material for clothing and household fabrics
Services	Flood and storm protection	Carbon sequestration
	Sediment retention	Climate regulation
	Shoreline stabilization	Aquifer recharge and discharge
	Water quality	Soil erosion control
	Aquifer discharge and recharge	Storage and recycling of nutrients
	Fauna and flora conservation	Biological control
	Transportation	Maintenance of biological diversity
	Research and training opportunities	Scientific and educational opportunities
	Support to external economic activities	
Attributes	Biodiversity conservation	Aesthetic information
	Existence value (ethic and moral values)	Cultural inspiration

Source: Windevoxhel (1992) and adapted from Sención (1996) and De Groot (1992).

over time because of the lack of information about different management scenarios. As a final step, the stock values were quantified according to the species harvesting abundance, and annual flow values of harvestable goods were quantified according to their production cycles.

The economic value of goods in stock corresponds to the potential gross income generated from the existing harvestable biomass in the forest, meaning that natural capital is evaluated in terms of its productive capacity. The value of economic goods flows is measured according to the extraction rate of the resources, the associated costs and the economic price of the products. Direct economic costs include input, labour and transportation costs to the export frontier.

With respect to the available information, it is important to note the following limitations.

1. Little information was available for selecting and valuing ecological functions in Nicaragua. In particular, the major part of productive and ecological information was selected from the literature. Primary information generated by the Olafo project was the main input used for the evaluation of goods. The lack of hydrological information impedes the valuation of important ecological functions, such as sediment retention, which are characteristic of mangroves ecosystems,
2. The majority of data used for the harvestable forest products in Guatemala was generated by project research, including preliminary results on timber and non-timber resource growth rates. In addition, primary information on production costs and prices comes from systematic field research on forest management activities and surveys of private producers. However, the lack of ecological/productive information over time on

Table 7.2 Available information on studied areas (mangroves in Nicaragua, broadleaf forest in Petén, Guatemala)

Mangroves of the Pacific Coast, Nicaragua	Broadleaf Forest, Petén, Guatemala
Productive and financial evaluation of production systems	Forest inventories
Financial evaluation of productive alternatives	Preliminary research results: growth rate, harvestable products, rotation cycle
Topographic maps and aerial photos	Financial evaluation of timber and non-timber extraction/processing
Extensive field contact in the area (local populations, national institutions, commercial sector)	Extensive field contact in the area (local population, national institutions, commercial sector)
	Preliminary productive and financial evaluation of production systems

Source: Windevoxhel (1992) and Sención (1996).

non-commercial timber species and non-timber species (data quality) has limited the scope of the study. In both cases, the studies were limited to one year. The available information for the study areas is described in Table 7.2.

Physical and Economic Valuation Methods: the Case of the Mangroves in the Pacific Coast of Nicaragua

Delimitation of the study area, identification and selection of goods, services and attributes

The boundary of the study site was defined according to the mangrove vegetation limits and to adjacent tidal areas. From the goods and services commonly found in mangroves ecosystems (Maltby, 1986; Dixon and Sherman, 1990; Pearce and Turner, 1990; McNeely, 1988; Dugan, 1990, cited in Windevoxhel, 1992), six goods and two services were valued based on those identified in the study area through field studies and on available field and secondary information (Table 7.3).

Table 7.3 Identified and selected goods and services in the mangroves of the Pacific Coast in Nicaragua

	Identified in the Study Area	Selected and Valued
Goods	Forest products (fuelwood, poles, bark for tannin)	Forest products (fuelwood, poles, bark for tannin*)
	Sea and estuarine fishing	Estuarine fishing
	Shellfish extraction (*Anadara tuberculosa*, *Ucydes occidentalis*)	Shellfish extraction (*Anadara tuberculosa*, *Ucydes occidentalis*)
	Shrimp extraction	
Services	Recreation	Recreation
	External support: industrial shrimp fishing	External support: industrial shrimp fishing
	Storm protection	
	Nutrient retention	
	Sediment retention	
	Flood control	

Note: * Valued only for the sustainable scenario.

Source: *Revista Forestal Centroamericana*, 9(3), 1994.

In the traditional scenario, bark production, shrimp extraction and sea fishing were not selected because the first two were not economically important within the local and national economy and, for the third, the relation between the productive and reproductive cycles of the most commonly extracted species and the mangroves was not established. Wasting wood bark was identified as a limitation in the traditional system. Fuelwood production processing involved the following activities: cutting the wood, debarking the trunks and leaving behind the potential raw material to produce tannin. In the sustainable scenario, the use of bark for tannin was proposed as a way to improve the traditional production system.

Flood control and nutrient and sediment retention were not selected for either scenario because of the lack of information on hydrological characteristics and on contamination inputs and outputs with respect to the mangroves.

The two services selected were recreation and mangroves' support to external productive systems. The studied area has a high flow of regional/national tourists due to the existence of beaches, estuaries and mangrove vegetation easily accessible from Managua and other cities in Nicaragua. In the present study, no difference was made between tourism (activity lasting for more than 24 hours) and recreation (activity carried out for a shorter period). With respect to external support, the commercial shrimp fishing industry represents an important source of export revenue to the national economy. The importance of the contribution of mangroves to this activity is due to the dependency widely recognized in the literature (D'Croz and Kwiecinski, 1980 cited in Windevoxhel, 1992) of commercial crustacea upon mangroves and coastal lagoons during their juvenile life cycle.

Goods

Forest productivity was defined according to Saenger (1989) who reported a total net productivity of 6.84 g/m^2/day (average of ligneous and leaf parts) in the mangroves of Nicaragua (Windevoxhel, 1992). According to Prado and Silva (1992), only 24.8 per cent of the total biomass is harvested. Based on the net primary productivity and the dry density factor of the wood, potential annual fuelwod production was estimated at approximately 19 600m^3 in the study area of 8700km^2. Total annual wood consumption was estimated at about 26 600m^3, indicating approximately 7000m^3 of overexploitation. The estimated wood demand does not include poles because demand is seasonal and no estimation was available at this time. Assuming that the forest is clearcut, it represents an annual loss of approximately 400 hectares of mangroves. At this rate, mangroves would disappear within seven to 10 years.

Due to the lack of reliable biological information, it was not possible to evaluate the sustainable physical limit for fishing and shellfish. Based on direct observation (CATIE/IUCN, 1991) it was determined that the current captured product size was less than the expected sizes reported in the literature. As a result, it is assumed that the extraction level for these goods is at its maximum (Windevoxhel, 1992).

Labour costs represent more than 85 per cent of total costs in the production/extraction systems, demonstrating the importance of manual labour in productive activities in comparison with capital-intensive activities. Due to unemployment, the shadow price of labour for all activities was conservatively estimated at 75 per cent of the market price.

Services

Recreation Recreational services were valued with two alternate methods: the travel cost method (TCM) and tourists' willingness to pay for the conservation of the mangroves. The travel cost is estimated as fee (in this case, equal to zero) plus the travel cost plus the opportunity cost of free time.

Total annual demand was valued as the sum of the average visitor's travel cost multiplied by the number of trips made by each person. A modification was introduced to the model commonly used in the industrialized economies where the opportunity cost of free time ranges between 25 and 50 per cent of the market value of the daily wage earned (Rosenthal *et al.*, 1986, cited in *Revista Forestal Centroaméricana*, 9(3), 1994). In lesser developed countries such as Nicaragua, there is a wide wage gap, where the majority of the population earns a very low salary and a minority earns high wages. The opportunity cost of free time for the higher-wage class was set at 50 per cent of the total wage, and increased to 85 per cent for the wage class in a more critical economic situation.

In addition to this method, the contingent valuation method was applied to compare the results obtained with the TCM.

Support to external economy The service provided by the mangroves to industrial shrimp fishing in open sea was measured according to a simulation model which establishes a relationship between shrimp capture and the mangrove area (surface). The simulation model was used by Lynne *et al.* (1981) to evaluate blue crab production in the Florida coast:

$$C_t = B_0 + B_1 (M_{t-1})^{0.5} (E_t) - B_2 (M_{t-1})^{0.5} (E^2_t) + B_3 C_{t-1} + e \qquad (7.1)$$

in which:

C_t = shrimp capture (pounds of tails),
M = forest mangrove area (in hectares),
E = fishing effort (number of shrimp boats),

e = error, and
B_0, B_1, B_2, B_3 are statistical parameters.

In order to estimate the variation in shrimp harvesting through time, E was set as a constant and M was reduced annually by 1000 hectares, according to the tendency registered in aerial maps for the period 1971–89 (Windevoxhel, 1992). The economic valuation of projected commercial shrimp harvest was based on total and per hectare net income generated by the activity. Production costs and international prices of the product were defined according to the prevalent situation of the commercial shrimp firms and to statistics from the governmental institution in charge of fishing control.

The different scenarios of mangroves resource management Management options can have two objectives: to maintain or recuperate ecosystem quality or to offer management alternatives for the system which, in addition, allow for increasing economic welfare of local population (Windevoxhel, 1992).

Two basic scenarios were evaluated: in the first, the extraction practices continue according to current tendencies and, in the second, forest resource extraction is not higher than the estimated net productivity defined by Saenger (1989). Due to the importance of forests for the conservation/reproduction of the associated shellfish resources in the ecosystem, and because of the lack of information on shellfish and fish productivity, the basic assumption is that sustainable use or overexploitation of forest goods directly affect the services provided by the ecosystem.

In the sustainable scenario, improved fuelwood management practices and the production of tannins for bark were introduced and valued. These improved management practices are supposed to have a direct positive effect on environmental services (in this case, the support to external activities) and on the recreation services due to a stabilization, over time, of the mangroves forest quality and area.

The basic factors taken into account in the sustainable scenario include the following.

1. Wood extraction including roots and branches with a diameter greater than 4 cm which are traditionally not harvested. Total commercial biomass estimated with these parameters increases from 24.8 per cent (in the traditional management practice) to 44 per cent of total net productivity (Prado and Silva, 1992, cited in Windevoxhel, 1992). The application of this practice would generate a production equivalent to 34 736m^3, without affecting the forest, which is more than current local wood demand (26 623m^3).
2. Although bark resources represent 5.6 per cent of total mangrove wood

biomass, the harvested fuelwood logs contain only about 53 per cent of total tree bark available. Assuming a 75 per cent efficiency in bark extraction for tannin elaboration, the total value of potential production of goods was calculated based on the substitution cost of importing 'quebracho' from Argentina. The replacement of import products with domestic goods implies hard currency saving for the Nicaraguan economy.
3. Shrimp production was considered as a constant due to the fact that wood management will be sustainable, so that mangroves will not be affected.

Sensitivity analyses (the third and fourth scenarios) were elaborated considering the following assumptions:

- a decrease of 40 per cent in the price of the shrimp export products due to external factors (such as cholera, an epidemic directly affecting Nicaragua's fishing sector);
- a full employment situation which sets the labour opportunity cost at the same level as the market price of labour.

The projection period was set at 10 years because at current rate of fuelwood extraction it is expected that the forest mangrove will disappear after seven to 10 years. The Central Bank of Nicaragua did not set a specific social discount rate and, for the evaluation, this later was set at 10 per cent, following Lutz and Munasinghe (1991).

Physical and Economic Valuation Methods: the Case of Tropical Forest in Guatemala

Study site delimitation and identified goods, services and attributes in the study area

The study area was delimited according to forest area defined in the San Miguel Community Concession contract, and covers about 6500ha (Seneión, 1996). Considering the relatively homogeneous physical characteristics of the Concession area in terms of slopes and soil composition, no significant impacts from external agricultural areas on the forest functions were identified. Based on goods, services and attributes commonly found in this type of ecosystem, nine goods, five environmental services and one attribute were identified, of which two goods, three ecological functions and the attribute were selected for valuation (see Table 7.4).

Potential commercial timber species were not selected because of the lack of both ecological/productive and economic information. As explained in the methodology, alternative management scenarios were evaluated. Instead, one

Table 7.4 Identified and selected goods, services and attributes in the broadleaf forest of the San Miguel Community Concession

	Identified in the Study Area	Selected and Valued
Goods	*Cedrela odorata*	*Cedrela odorata*
	Swietenia macrophyla	*Swietenia macrophyla*
	Potential commercial timber species[2]	
	Chamadorea	
	Manilkara	
	Desmoncus	
	Pimenta dioica	
	Hunting	
	Honey production	
Services	Tourism	Carbon sequestration
	Carbon sequestration	Nutrient cycling
	Nutrient cycling	Erosion control
	Erosion control	
	Aquifer recharge	
Attributes	Biodiversity	Biodiversity

Source: Seción (1996).

management type (the extraction based on harvestable possibility) was taken into consideration to value both forest stock and flows. Given topographic, hydrological and soil characteristics – in particular for calcareous soils – the calculation of aquifer recharge effects of the forest was not possible. Tourism was not selected as this activity is not economically important in the study area.

Carbon sequestration, nutrient cycle and erosion control environmental services were selected because of the important role they play in this type of ecosystem (De Groot, 1992; Vitousek and Sanford, 1986, cited in Sención, 1996) and because valuation was feasible with the available secondary and primary information. The biodiversity attribute was selected because of the importance of the Biosphere Reserve, not only at national level but also at international level. Although this non-use value is not easy to quantify, international and national organizations' willingness to pay for the conservation and management of the Reserve can be calculated.

Goods

Cedrela odorata and *Swietenia macrophyla* Research results on the abundance of harvestable timber (CONAP, 1995) in a forest similar to San Miguel show 0.2 individuals per hectare for Cedrela and Swietenia. Total physical stock which corresponds to total harvestable volume per hectare for both species with a diameter at least equal to 60cm is $0.59m^3$/ha (132.16 bf/ha) (Sención, 1996). According to Galvez (1996) cited in Sención (1996), preliminary results of annual growth rates range from 0.7 to 2.1mm for *Cedrela odorata*, and 1.4 to 4.2mm for *Swietenia macrophyla.* Consequently, the harvesting cycle for both species is estimated at 40 years (Galvez, 1996, cited in Sención, 1996) and the annual harvestable flow is 3.30 board foot per hectare per year. Timber free on board (FOB) price (US$0.86/bf), extraction and processing cost in the forest (US$0.08/bf) and transportation cost (US$0.17/bf) were based on primary data.

Chamadorea spp. Xate *(Chamadorea spp.)* leaf stock is estimated at 457 harvestable leaves per hectare (Galvez, 1996, cited in Sención, 1996). Production is estimated at 2.28 leaves per plant per year. Consequently, annual physical flow of xate is 879 leaves/hectare. The FOB price was estimated according to official information from export firms and the Bank of Guatemala. Field data from the Olafo project estimate the extraction and first processing cost at US$1.45/ha/year according to labour productivity of local 'farmers'. Transportation, product selection, packaging and storage costs at the warehouse level were not available. As a result, total net economic income of *Chamadorea* is overestimated.

Manilkara achras Due to the lack of specific information from the study area, the harvestable quantity of latex was estimated at 0.46kg/tree according to Reining and Heinzman (1992). The abundance of Manilkara in the adjacent area of La Pasadita was set at 9.4 trees/ha. Consequently, total physical stock is 4.32 kg/ha. The rotation cycle used in Petén is equal to five years, and annual flow per hectare is approximately 0.86kg. Latex extraction, transformation and commercialization processes in Petén have been traditionally organized by a syndicate. As a result, a series of export prices for the period 1981–94 are available at the Central Bank of Guatemala. Direct extraction costs were estimated based on the labour productivity (3.9kg of latex per day) and to the labour opportunity cost in the area (Reining and Heinzman, 1992).

Desmoncus sp. The *Desmoncus sp.*, a climbing fibre palm used for handicrafts called 'bayal' is abundant in Petén and underutilized by the local population. Research on ecological, productive, processing and economic aspects allow for physical evaluation as well as economic valuation of this resource directed towards handicraft products elaboration. According to Chinchilla (1994), cited in Sención (1996), average product harvested for each stem was 95.13m. Considering that the minimum cycle rotation is four years, physical stock of fibre for handicraft was estimated at 381m/ha and annual flow at 95.25m/ha (Galvez, 1996, cited in Sención, 1996). Because this product is not exported, the *Salix sp.* which is currently marketed for its fibre was considered as a substitute product. Direct labour costs for extraction and transformation of raw material were available from the project's field research.

Services

Carbon sequestration In order to avoid double accounting with the erosion control service, carbon included in the soil and roots was not included; instead, this element was quantified for the aerial dry biomass. A conversion factor from dry biomass to carbon was set at 50 per cent (Brown and Lugo, 1980, cited in Sención, 1996) and the biomass expansion factor at 1.6 (Lugo and Brown, 1984, cited in Sención, 1996). Physical stock of carbon was calculated according to the following equation (Sención, 1996):

$$PSc = V \times EFB \times D \times CFC \qquad (7.2)$$

in which:

PSc = physical stock of carbon (t/ha),
V = bole volume (m^3/ha),
EFB = biomass expansion factor (1.6),
D = wood density (0.62t/m^3),
CFC = biomass conversion factor to carbon (0.5).

Annual flows correspond to the amount of carbon sequestered annually and were estimated according to the annual average increase of basal area (m^2/ha/year) taking into account the different species and diameter classes (Galvez, 1996, cited in Sención, 1996), as shown in the following equation (Sención, 1996).

$$PFc = MAI \times CFV \times D \times CFC \qquad (7.3)$$

in which:

PFc = physical carbon flow,
MAI = mean annual increment (m^2/ha/year),

CFV = volume conversion factor as the arithmetic quotient between volume and basal area of different diameter classes,
D = wood density (0.62 t/m^2),
CFC = biomass conversion factor to carbon (0.5).

The economic price of sequestered carbon was selected based on the different prices paid by international organizations which support reforestation projects and is equivalent to a conservative price of US$5.5/t (Dixon *et al.*, 1993, cited in Sención, 1996).

Erosion control Physical soil stock was evaluated according to soil characteristics and slopes; soil flow, equivalent to erosion soil loss, was calculated with the USLE (universal soil loss equation) considering only erosion due to precipitation (output) and not soil formation processes due to the lack of information.

The basic equation used for soil stock definition is:

$$SS = DS \times E \times Sd \tag{7.4}$$

in which:

SS = soil stock,
DS = depth soil (0.20m),
E = 10 000m^2 (1 ha),
Sd = soil density (t/m^3).

The universal soil loss equation was used:

$$SL = R * K * LS * C \tag{7.5}$$

in which:

SL = soil loss,
R = erosive precipitation factor,
K = erosive soil factor,
LS = slope length and gradient,
C = vegetation cover factor.

According to the parameters available for San Miguel, the soil loss was estimated at 2632ton/ha.

The erosion control function was evaluated using two alternative methods: the replacement cost which allows for quantifying the nutrient loss through soil loss calculated with the USLE; and the preventive cost method to measure soil conservation practice costs. In the first case, the costs include:

- soil commercial value adjusted by the standard conversion factor to transform market prices to shadow prices for those products not marketed at international level (Squire and Van der Tak, 1980);
- transportation costs of soil to the study area.

In the second case, different systems for preventing erosion were considered (stone retention fences, drainage ditches, terrace fields, home garden patches) and their implementation was evaluated with economic labour costs. The costs were used to value physical stock (t/ha) and soil loss flows (t/ha). Annual flows were valued in terms of indirect costs and not through indirect benefits, while stock values were based on the indirect benefit provided by the forest vegetation.

Nutrient cycling The nutrient cycling stock was evaluated according to the proportion of nutrients (nitrogen (N), phosphorous (P) and potassium (K)) in the dry aerial biomass:

$$PSn = V \times EFB \times D \times NPKP \tag{7.6}$$

in which:

PSn = physical stock of nutrients ($t_{N,P,K}$/ha),
V = bole volume (m^3/ha),
EFB = biomass expansion factor (1.6),
D = wood density (0.62t/m^3),
$NPKP$ = N, P, K proportion in the dry biomass ($T_{N,P,K}/T$).

For an average of 9.65 t/ha of humus dry biomass, nutrients were estimated at 0.01787 t_{NPK}/t of dry biomass. This proportion of NPK for each ton was applied for both nutrient stock and flow.

To estimate annual flows, the proportion of nutrients in the humus dry biomass was considered:

$$PF_{NPK} = HDB \times P_{NPK} \tag{7.7}$$

in which:

PF_{NPK} = physical nutrient flows (N,P,K),
HDB = humus dry biomass,

P_{NPK} = proportion of N, P, K in the dry biomass (T_{NPK}/T).

For the economic evaluation of nutrient cycling stock and flows, the economic price of fertilizer was set according to physical estimation of N,P,K (the proportion of NPK for each ton of fertilizer was set at 45 per cent).

Attributes

Biodiversity The biodiversity attribute was valued indirectly based on the support given by international organizations for conserving the Mayan Biosphere Reserve. The total (stock) and annual (flow) value of funding directed to the MBR was used (and adjusted on a per hectare basis) to approximate the national and international communities' willingness to pay to conserve the tropical biodiversity of the area. Four funding sources were considered: the Agency for International Development from the USA (AID), the Danish International Development Assistance Agency (DANIDA), the Norwegian Agency for Development (NORAD) and the Swedish International Development Agency (SIDA).

Assumptions The following assumptions were made for the economic evaluation of the forest ecosystem:

- timber extraction is carried out according to traditional practices and includes two precious timber species (*Cedrela odorata* and *Swietenia macrophyla*) with a DBH (diameter breast height) measurement higher than 60cm;
- lesser used species are not extracted because of the lack of market;
- non-timber extraction is carried out according to traditional practices and refers to two species (an ornamental plant, *Chamadorea spp.*, and a latex, *Manilkara achras*);
- a new non-timber species, *Desmoncus sp.*, has been identified due to its potential commercial value.

RESULTS

Value of the Mangroves of the Pacific Coast of Nicaragua

In the 8700ha considered, net benefits of traditional management of natural resources is US$5.7 million over 10 years, considering a discount rate of 10 per cent. In the second (sustainable) scenario, net benefits are almost twice those in the traditional scenario (US$11 million), demonstrating the higher

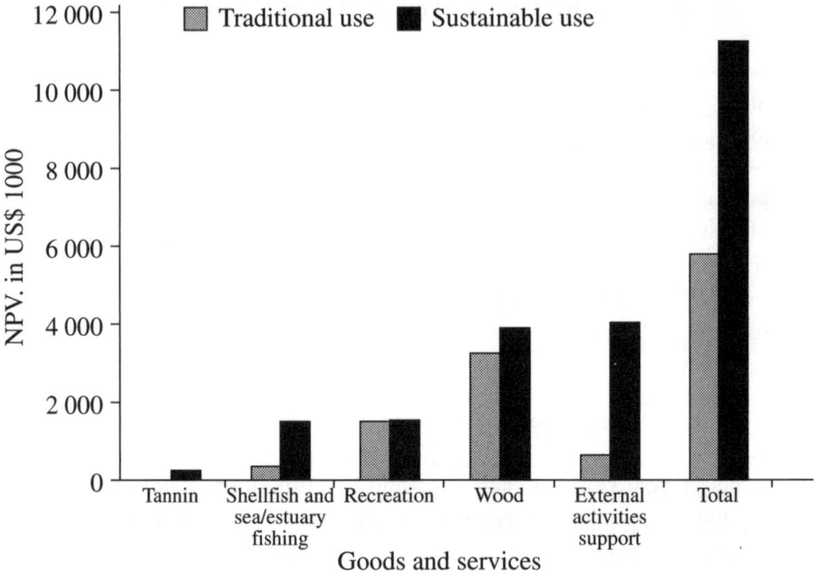

Figure 7.5 Goods, services and total ecosystem values: mangroves of the Pacific coast in Nicaragua, scenario 1 (traditional use) and scenario 2 (sustainable use)

value provided by ecosystems managed in a more sustainable way (see Figure 7.5). The ecosystem value on a per hectare basis, generated by goods and services is US$700 in the traditional scenario and US$1300 in the sustainable scenario.

An analysis of the services valued according to the methodology previously described shows that recreation services have a lower value using the contingent valuation (CV) than with the travel cost method (see Table 7.5). The lower willingness to pay is probably due to the critical national economic situation and the tourists' reluctance to pay for public-access goods.

The mangrove support to industrial shrimp fishing represents US$77 per hectare, higher than the value reported in existing literature (Costanza *et al.*, 1989; Lynne *et al.*, 1981, cited in Windevoxhel, 1992). This situation can be attributed to economic factors such as the high added value of exported shrimp and relatively low production cost levels, or to biological factors such as the differences in the mangroves' productivity (Windevoxhel, 1992).

If current extraction of woody products continues at the same rate, it is expected that in year nine the mangrove will disappear. The associated loss in coastal lagoon quality would imply a decrease in the recreation activities on the part of the tourists who visit the estuaries (5 per cent of total) and of those

Table 7.5 Total and average travel cost (US$) according to the two valuation methods used

Valuation method	Monthly average (US$)	CV %	Annual average (US$)	Total annual (US$)
Contingent	4.44	117+	53.30	9 038.4
Travel cost	—	591*	175.56	252 616.9

Note: +: of monthly average; *: of annual average; n = 680.

Source: Windevoxhel (1992).

who visit the beaches, estuaries and mangroves during the same trip (44 per cent).

The net present value of each good and function in the traditional (1) and sustainable (2) scenario is reported in Figure 7.5. In the traditional scenario, the large benefits obtained from fuelwood (US$3 249 967, or about 90 per cent of value generated by all the goods) are mainly due to the low opportunity cost of labour, implying very low production costs. As a result, goods had a higher value (US$3 596 785) when compared with services (US$2 143 365). In the second scenario, the goods have a value almost equal to the services value. This is due to the fact that all the goods were valued, as well as including an additional good (bark for tannin).

This result confirms the importance of goods generated by ecosystems in the American wetlands (Saenger, 1989; Snedaker, 1986, cited in Windevoxhel, 1992). In this particular case, the dependency of the local population on extractive resources has been demonstrated. On the other hand, tourism potential in the area is underexploited due to the lack of infrastructure and incentives. In addition, the relatively lower value of services is due to the fact that at least two ecological functions identified were not valued because of the lack of information. For example, Costanza *et al.* (1989) reported that flood protection of the mangroves of Louisiana in the United States, valued by the contingent valuation method, contributed 78 per cent of total benefits (Windevoxhel, 1992).

Sensitivity analysis of labour price increases (scenario 3) demonstrates a significant decrease of the contribution of goods to total value (see Figure 7.6): 37 per cent of total economic value of the ecosystem instead of 48 per cent in the second scenario. This is indicative of the strong incentive to extractive activities in depressed economies. Nevertheless, this scenario has a higher value than the traditional one (US$8 826 155 and US$5 740 150, respectively). In the fourth, and most pessimistic, scenario (Figure 7.6)

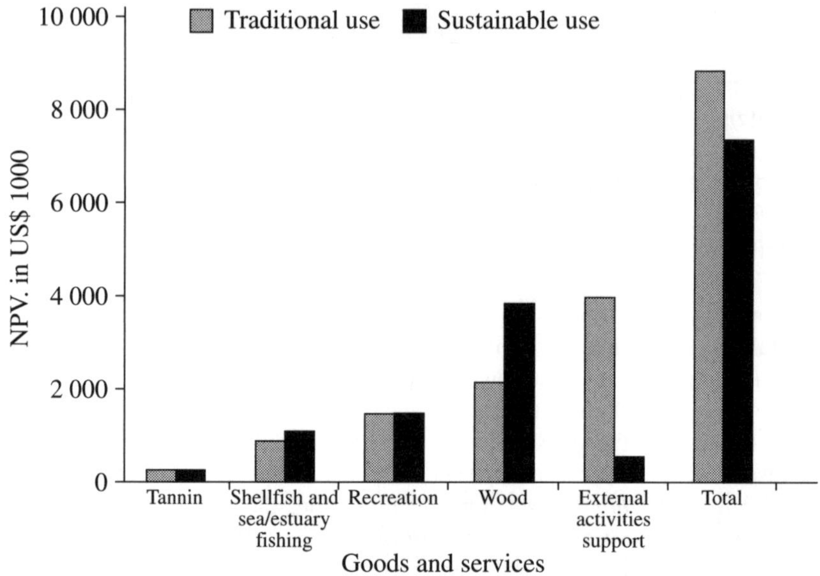

Figure 7.6 *Goods, services and total ecosystem values: mangroves of the Pacific coast in Nicaragua, sensivity analysis scenarios 3 and 4 (increase in labour price vs. decrease in goods prices)*

which considers lower prices of goods during the entire period considered, the total ecosystem value continues to be higher than in the traditional scenario.

From the results presented above, we can confirm that the economic valuation of mangroves based on different types of management practices is feasible. However, there is a need to adjust traditional valuation methods to specific economic conditions of developing countries. This is the case in particular for the TCM, for which adjustments were made considering the high level of underemployment, the low level of satisfaction of basic needs, the low importance given to free time compared with the developed countries, the importance of public transportation versus individual (particular) transportation and, finally, the unequal income distribution.

This type of evaluation demonstrates in precise terms the pressure exerted on mangroves resources by the local population. Due to the fact that crucial environmental services linked to hydrological process were not included, the final evaluation is conservative. However, study results demonstrate the dependence between the management of goods and the conservation of services. In this particular case, services and extractive goods have a complementary contribution to the economic value of the ecosystem.

Economic Value of the Forest in Petén, Guatemala

The stock value generated by the forest is US$5 072 850, equivalent to US$787/ha, while annual flow is US$195 543, equivalent to US$30/ha. The flow value is within the range of 24 to 72 cited by Carranza *et al.* (1996) (US$24 and US$72 per hectare). Figures 7.7 and 7.8 show stock and flow value, respectively, for each good, service and attribute.

With both stock and flow values, environmental services contribute to the greater part of total value (80 per cent and 62 per cent, respectively). The erosion control contributes with 73 per cent of total stock value of environmental services, while nutrient cycling is predominant (62 per cent) when evaluating the annual forest flow values.

The value of traditional timber species (*Cedrela* and *Swietenia*) represents 75 per cent of total stock value of the goods while non-timber products represent only 12 per cent, 10 per cent and 2 per cent for *Desmoncus*, *Manilkara* and *Chamadorea*. On the flow side, the non-timber products contribute more (80 per cent) than timber to the total. This is due to the shorter harvesting cycle of non-timbers products and to the low abundancy of valuable timber species. Also this situation demonstrates that the price of

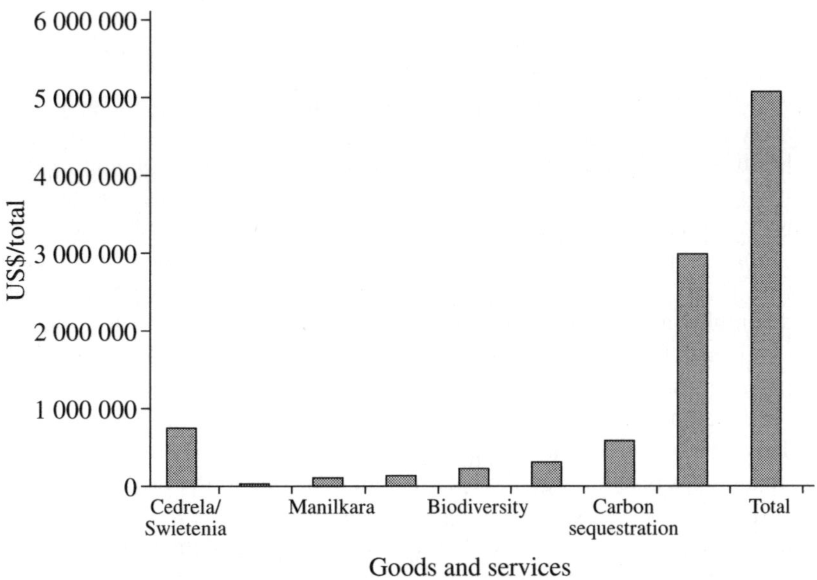

Figure 7.7 Subtropical forest of San Miguel, Petén, Guatemala: stock values of goods, services and total ecosystem

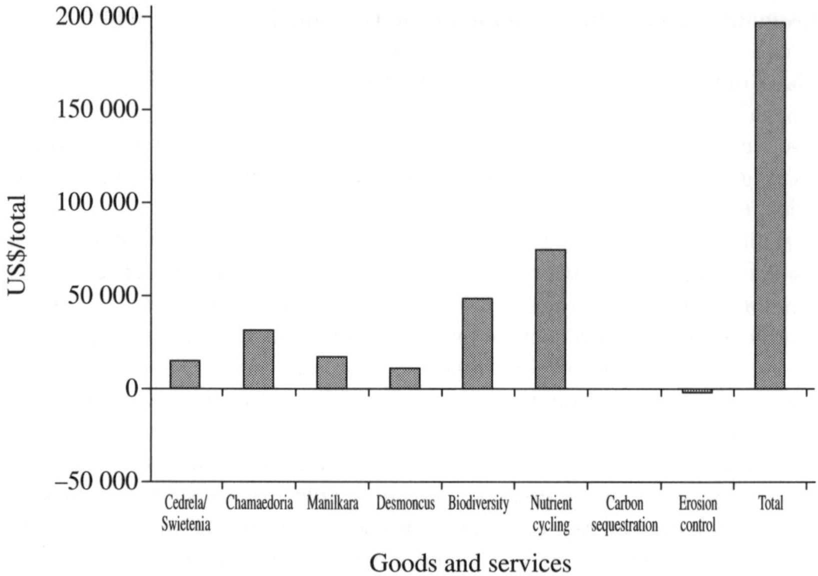

Figure 7.8 Subtropical forest of San Miguel, Petén, Guatemala: flow values of goods, services and total ecosystem

precious timbers does not compensate for low abundance and productivity, the elements which define the harvesting rotation cycle.

With respect to environmental services, the erosion control function generates 59 per cent of total stock value of ecological functions and attributes, while carbon sequestration contributes 11 per cent and the remaining functions less than 6 per cent each. The importance of the erosion control function is probably due to poor soil characteristics and the valuation method. As seen in the description of the study site, soils are predominantly calcareous, shallow and, without a forest cover, they are easily dispersed. Even using the opportunity cost of labour which represents 43 per cent of the market labour price, the high stock value can be attributed to the high number of days (192 days/ha) required to implement soil conservation practices.

Of the total annual flow values generated by environmental services and goods, the nutrient cycle function and, to a lesser extent, the bioversity attribute contribute with 62 and 39 per cent, respectively. The importance of the nutrient cycle service reflects the low soil fertility and the role of fallen leaves as a main source of nutrients. In effect, in San Miguel, crop areas are abandoned to fallow after three to five years.

The negative contribution of the erosion control function in the annual flow is due to the fact that no positive annual flows were measured for soil formation; the erosion control was valued only through its indirect costs (see above). It is important to note the low value of carbon sequestration due to the slow tree growth rates and the inclusion of only the commercial species.

In conclusion, given the ecological and economic context of this forest, ecological functions and attributes generate more value when compared with the goods. The use of indirect valuation methods to quantify ecological functions and attributes may led to a final overestimation because, according to the available information, all function values were defined, except for the erosion control, according to their indirect benefits. The contributions of goods and functions reflect the importance of ecological functions in tropical forests. In particular, the low carbon sequestration value can be explained by the fact that, in a mature forest, flows are relatively low when valued over the long run.

The comparatively low value of goods is due to the lack of commercial opportunities for more predominant non-traditional timber species and the overexploitation of the two valuable species. The results show that diversified management, introducing the sustainable use of previously unexploited non-timber products, increases the forest's economic value.

Two important factors affect total ecosystem value of the ecosystem: first, the traditional forest management is labour-intensive, and second, the depressed economic conditions of the region imply, among other things, a low labour opportunity cost. These elements are crucial and should be considered in a sensitivity analysis of the forest value.

From a methodological perspective, the valuation based on stock and flows allows for a better understanding of the dependence between natural capital and annual 'interest' of the forest, according to specific management practices. Based on this evaluation, specific research topics and decisions oriented towards the sustainable use of the forest can be identified. In particular, the need to diversify the marketable timber species, and to promote a sustainable management of non-timber products as complementary activities, is evident. Such actions will not only generate more commercial goods but will contribute also to the more efficient use of ecological functions such as carbon sequestration.

CONCLUSIONS

Conclusions from the two case studies refer to two main aspects: methodology and ecosystem values. The general guidelines and the specific methodology applied in each case show that the economic valuation of ecosystems is feasible. Furthermore, the existing valuation methods can be used/adjusted to the particular economic conditions of developing countries. Despite the differences in the way each ecosystem was valued (projection over time according to net productivity of the mangroves and stock and flows of the broadleaf forest), both took into account the type of natural resource management. The evaluation in terms of stock and flows allows for a reconciliation between ecology and economics: it becomes clearer that natural capital (stocks) measured according to harvestable potential generates flows (interest) which depend upon management practices.

Another important issue is the demonstration of the relationship between extraction/management of goods and environmental services. In the case of the mangroves, the sustainable use of forest goods allows for the conservation of ecological functions and services (in this case, support to commercial shrimp fishing and recreation). Beyond the results obtained, the methodology used allows for the identification of missing ecological and economic information required for a more reliable partial evaluation of the ecosystem. In this sense, it is a tool for research organizations' decision makers.

Regarding the results on economic valuation, the three hypotheses (the sustainable management of these forests provides a higher value than traditional extractive practices; the sustainable management of goods implies that ecological functions create more value; ecological functions provide more value than goods) were verified in the mangroves case study. In the case of the broadleaf forest, the comparative analysis of different scenarios was not elaborated and consequently only the third hypothesis was positively verified.

No comparison in the results can be made between the two ecosystems due to the different way of valuing them. Both areas are characterized by a marginal economic situation, an overexploited level of commercial products, limiting ecological factors and the production of goods with low added value. All these factors mean that, in the short term, environmental services can be affected negatively if no actions are taken. In particular, actions should be directed towards the improvement of the management of traditional resources/species and, simultaneously, to the promotion of products which currently do not have commercial value. This strategy is very important if considering, as in the case of the mangroves, the sensitivity of the products' economic values to external factors such as climate, and labour price in particular.

One of the main limitations of the applied methodology consists of the direct dependence between the ecosystem value and its use intensity. In effect, if export goods are valued instead of local/national consumptive goods, the result will vary significantly. It is important to clarify that the economic valuation of natural resources does not provide a solution to the natural resource conservation and management decisions; it is rather a technical tool which can be used to compare different scenarios in order to provide elements for decision making.

NOTES

1. *Pouteria durlandii, Brosimum costaricanum, Dendropanax arboreus, Pseudolmedia spuria, Sebastiana longicuspis, Spondias mombin, Manilkara zapota, Pouteria campechiana, Vitex guameri* and *Pouteria amygdalina.*
2. *Pseudobombax ellipticum, Pithecelobium arboreum, Vatairea lundelli, Astronium graveolens, Aspidosperma megalocarpum, Pseudolmedia oxiphyllaria, Calophyllumm brasiliense, Longocarpus castilloi.*

REFERENCES

Barbier, E.B., R. Costanza and R. Twilley in collaboration with Tania Ammour *et al.* (1993), 'Guidelines for Tropical Wetland Evaluation', Technical Report no. 231, CATIE, Proyecto Conservacíon para el Desarrollo Sosterrible en América, Turrialba, Costa Rica.

Barton, D.N. (1995), 'Valoración económica parcial de alternativas de manejo para los humedales de Térraba Sierpe, Costa Rica', MSc. thesis, Centro Internacional en Política Económica, Universidad de Heredia, Costa Rica.

Brown, S. and A. Lugo (1980), 'Tropical secondary forests', *Journal of Tropical Ecology*, 6, 1–31.

Carranza, C., B. Aylward, J. Echeverria, J. Tosi and R. Mejias (1996), 'Valoración de los servicios ambientales de los bosques de Costa Rica', Centro Científico Tropical, ODA-MINAE, San José, Costa Rica.

CATIE/IUCN (1991), 'Proyecto de producción de leña en las zonas de amortiguamiento de manglares de la Región II – Nicaragua'.

Chinchilla, M. (1994), 'Caracterización de las poblaciones de Bayal (Desmoncus, spp.) con fines de aprovechamiento artesanal en la unidad de manejo forestal de San Miguel, San Andrés, Petén, Guatemala', agronomical engineering thesis, Institute of Agronomical Research, University of San Carlos.

Collinet, J. (1997), 'Potencialidades y limitantes de algunos suelos en San Miguel La Palotada, Petén', Guatemala CATIE, Proyecto Conservación para el Desarrollo Sostenible en Centroamérica, technical series, technical report no.299, Turrialba, Costa Rica.

CONAP (1995), 'Plan de manejo forestal, Unidad de manejo La Pasadita, San Andrés, Petén, Guatemala', Borrador, CATIE.

Costanza, R. *et al.* (1989), 'Valuation and Management of Wetland Ecosystems', *Ecological Economics* (The Netherlands), 2, 57–75.

D'Croz, L. and B. Kwiecinski (1980), 'Contribución de los manglares a las pesquerías de la Bahía de Panamá', *Revista de Bilogía Tropical*, 28 (1), 13–29.

De Groot, R.S. (1992), *Functions of Nature. Evaluation of nature in environmental planning, management and decision making*, Groningen, The Netherlands: Wolters-Noordhoff.

Dixon, J. and P.B. Sherman (1990), *Economics of Protected Areas; a new look at benefits and costs*, Washington, DC: Island Press.

Dixon, J.A., R.A. Carpenter, A.F. Fallon, P.B. Sherman and S. Manopimoke (1988), *Economic analysis of the environmental impacts of development projects*, London: Earthscan Publications Ltd, in association with the Asian Development Bank, Manila.

Dixon, R.K., K.J. Andrasko, F.G. Sussman, M.A. Lavinson, M.C. Trexler and T.S. Vinson (1993), 'Forest sector carbon offset projects: near-term opportunities to mitigate greenhouse gas emissions', in *Water, Air and Soil Pollution*, 70, 561–77.
Dugan, P.J. (1990), 'Wetland Conservation: a review of current issues and required action', The World Conservation Union, Gland, Switzerland.
Furst, E., D.N. Barton, G. Jimenez (1996), 'Case study analysis on the application of environmental and natural resources valuation methodologies in developing countries and countries in transition to market economies', International Center in Economic Policy for Sustainable Development (CINPE), Universidad Nacional de Heredia (revised version).
Galvez, J. (1996), 'Elementos técnicos para el manejo forestal diversificado de bosques naturales tropicales en San Miguel, Petén, Guatemala', Msc thesis, CATIE, Turrialba, Costa Rica.
IICA/FLACSO (1990), 'Centroamérica en Gráficas', Instituto Interamericano de Cooperación para la Agricultura–Facultad Latinoamericana de Ciencias Sociales, San José, Costa Rica.
Lugo, A. and S. Brown (1984), 'Biomass of tropical forests: A new estimate based on forest volumes', *Science*, 233, 1290–93.
Lugo, A.E. and S.C. Snedaker (1994), 'The ecology of mangroves', *Annual Review of Ecology and Systematics*, 5, 39–64.
Lutz, E. and M. Munasinghe (1991), 'Environmental–Economic Analysis of Projects and Policies for Sustainable Development', Environment Working Paper No.42, Environment Department, The World Bank.
Lynne, G. *et al.* (1981), 'Economic Evaluation of Marsh Areas for Marine Production Processes', *Journal of Environmental Economics and Management*, 8, 175–86.
Maltby E. (1986), 'Waterlogged Wealth: Why waste the world's wet places?', London and Washington, DC: International Institute for Environment and Development.
Marmillod, D., J. Galvez (1997), Efecto del aprovechamiento maderero sobre la población de bayal, una especies no maderable del bosque petenero: implicaciones para una silvicultura con fines de producción diversificada de este bosque', presented at the 'Simposio Internacional sobre posibilidades de manejo forestal sostenible en América Tropical', Sta. Cruz de la Sierra, Bolivia, 15–20 July.
McNeely, J.A. (1988), *Economics and Biological Diversity: developing and using economic incentives to conserve biological resources*, Gland, Switzerland: International Union for Conservation of Nature and Natural Resources.
Pearce, D.W. and R.K. Turner (1990), *Economics of Natural Resources and the Environment*, Baltimore: Johns Hopkins University Press.
Pearce, D., N. Adger, K. Brown, R. Cervigni and D. Moran (1993), 'Mexico forestry and conservation sector review. Substudy of economic valuation of forests', November 1993, Report to World Bank Latin America and the Caribbean - Country Department II (LA2), Centre for Social and Economic Research on the Global Environment, University of East Anglia and University College London.
Prado, M. and C. Silva (1992), 'Informe anual sobre vegetación: manglares Pacífico de Nicaragua', Universidad Autónoma de Nicaragua, León, Nicaragua, mimeo.
Reining, C. and R. Heinzman (1992), *Non timber forest products of the Maya Biosphere Reserve, Petén, Guatemala*, New York: Conservation International Foundation.
Reyes, R. (1996), 'Caracterización y Evaluación de la Sostenibilidad de los Sistemas de Producción de la Concesión Comunitaria de San Miguel. Petén, Guatemala',

Msc thesis, Centro Agronómico Tropical de Investigación y Enseñanza (CATIE), Turrialba.

Rojas, E. (1987), 'Estudio agroclimático para los departamentos León Chinandega', Región II de Nicaragua, Instituto Nicaragüense de Estudios Territoriales, Series IICA No. A1/NI-87-0001, San José, Costa Rica.

Rosenthal, D. *et al.* (1986), 'Users' Guide to RMTCM: Software for Travel Costa Analysis', Technical Report RM-132, Department of Agriculture, Forest Service, Fort Collins, Colorado.

Saenger, P. (1989), 'Economic evolution of tropical wetland resources', Project: Wide Use of Mangrove Resources in Estero Real (Nicaragua) and Terraloa-Sierpe (Costa Rica).

Sención, G. (1996), 'Valoración Económica de un Ecosistema Bosque Subtropical: Estudio de Caso San Miguel La Palotada, Petén, Guatemala', MSc thesis, Centro Agronómico Tropical de Investigación y Enseñanza (CATIE), Turrialba.

Snedaker, S.C. (1986), 'Aquaculture and Mangrove Ecosystem Productivity in Arid and Semi-arid Coastal Environments', *Marine Science of Arabian Sea. Proceedings of an International Conference, Keradi*, Washington, DC: American Institute of Biological Sciences.

Squire, L. and H. van der Tak (1980), *Análisis económico de proyectos*, Baltimore and London: The Johns Hopkins University Press: Madrid: Ed. Tecnos.

Vitousek, P. and R. Sanford (1986), 'Nutrient Cycling in Moist Tropical Forests', *Annual Review of Ecology and Systematics*, 17, 137–67.

Windevoxhel, N.J. (1992), 'Valoración económica parcial de los Manglares de la Región II de Nicaragua', Msc thesis, Turrialba, Costa Rica, CATIE.

Windevoxhel, N.J. (1994), 'Valoración económica de los manglares: demostrando la rentabilidad de su aprovechamiento sostenible. Caso Héroes y Mártires de Veracruz, Nicaragua', in *Revista Forestal Centroamericana*, 9, Año 3, September–November.

PART III

Forest Policy: Case Studies from North, Central and South America

8. The Canadian Model Forest Program: a case study of the Saskatchewan model forest[1]

Surendra N. Kulshreshtha

INTRODUCTION

The forest in many ways symbolizes Canada, from the earliest times when its inhabitants depended upon it for basic needs. Even today, forests cover half of Canada's landscape. Extending over some 453 million hectares of land, forests are a dominant feature of Canada's economy, its culture, traditions and history (CCFM, 1992, p. 1). The economic importance of this sector in Canada can be seen in terms of annual sales of $26 billion, and the employment of one out of 16 workers in Canada. In the international context, Canada also occupies a place of prominence, since almost 10 per cent of the world's forests are located here.

The Canadian Model Forest Program was designed to show leadership to the international community in matters related to forest management. As noted by Natural Resources Canada (1996, p. 7), 'the years leading up to 1993 had seen an increasing interest in the forests of the globe, perhaps best illustrated by the media coverage of issues related to forest management and the environment'. The federal government further noted that the eyes of the world were turning towards forest nations, including Canada, for charting the course of future forest management avenues. Among various initiatives that were undertaken, development of the Model Forest Program is perhaps the most significant, and also the one for which Canada is most renowned internationally.

The major objective of this chapter is to review Canada's Model Forest Program, with particular reference to a detailed study of one of the model forests, the Prince Albert Model Forest, in the province of Saskatchewan. This review is carried out with the following sub-objectives:

- to describe briefly the Canadian national and international Model Forest Program initiatives, particularly in terms of their objectives and scope, with special reference to the Prince Albert Model Forest;

- to assess the performance of the Prince Albert Model Forest Association in achieving the objectives of the Canadian Model Forest Program; and
- to identify the lessons that could be learnt from the experience of the Prince Albert Model Forest Association for future management of forests in Canada and elsewhere.

These objectives are achieved by a critical analysis of the Prince Albert Model Forest programme, using secondary data. Since the Prince Albert Model Forest is one of 10 such model forests in Canada, this analysis should not be construed as an assessment of the entire Canadian National or International Model Forest Initiative.[2] Furthermore, in this study, much of the analysis is carried out from a socioeconomic perspective. An assessment of the model forest's contribution in natural sciences areas, although desirable, is not attempted here. A further word of caution is advised: the Prince Albert Model Forest has only been in operation for a period of four years, including the initial year and half for establishment and planning. It is too short a period to achieve long-term objectives, and even to assess their efficacy.

The rest of this chapter is divided into five sections. The first section reviews the Canadian Model Forest Initiative and the Prince Albert Model Forest (PAMF) development, including its organization, activities and linkages with other model forest initiative activities. In the second section, the major thrust of the PAMF, including major issues facing it, is described. In the third section, an evaluation of the organization, including its success in meeting various stated objectives is presented. Lessons learnt from the PAMF experience are reported in the fourth section, which is followed by a summary and conclusions in the last section.

THE MODEL FOREST PROGRAM

Origin and Establishment of the Canadian Model Forest Program

Ever since the submission of the report by the World Commission on Environment and Development (WCED) in 1987, the attention of both public and private agencies has been brought to focus on sustainable development, which was also a part of the focus for the Model Forest Program Initiative. The origin of this initiative can be traced back to several concerns of the public, as noted by the Standing Committee on Natural Resources (see Nault, 1994, p. 1), particularly those related to ecosystem integrity and socioeconomic values, which were central to various concerns of policy makers.

The Model Forest Program was announced in September 1991 with the following objectives:[3]

1. to accelerate the implementation of sustainable development in the practice of forestry, in particular the concept of integrated resource management;
2. to apply new and innovative approaches, procedures, techniques and concepts in the management of forests; and
3. to test and demonstrate the best sustainable forestry practices utilizing the most advanced technology and forestry practices available.

Under the auspices of this programme, proposals were solicited for areas larger than 100 000 ha from groups that were committed to developing partnerships among various institutions and interest groups. The 10 selected Canadian model forests range in size from 108 000 ha to 1 500 000 ha, which reflect the scale at which forests are managed (see Appendix). One thing that was common to all successful proposals was the creation of partnerships. According to Brand *et al.* (1996, p. 74), these partnerships encompassed owners of the land, industry, First Nations, academic and educational organizations, a variety of municipal, provincial and federal agencies, and non-governmental organizations. However, in all cases the model forest was a new institution, while each of the members retained its original mandate.

Expansion to International Model Forests Network

Extension of the Canadian Model Forest Initiative to other countries started in 1993, when Mexico's two model forests joined the network. Since that time, another model forest has been accepted in the boreal forests of the far east of Russia. At this time, according to IMFN (1996) and Natural Resources Canada (1996), another six model forests are in the final stages of negotiations (or have joined recently).[4] Compared to the Canadian model forests, International Model Forests tend to be small in size, although they do represent a different forest region, as shown in Appendix A.

Formation of the Prince Albert Model Forest

The Prince Albert Model Forest (PAMF) joined the model forest network in April 1993. The partnership[5] included seven members: (1) federal government (through the Canadian Park Service); (2) provincial government (through the Saskatchewan Parks and Renewable Resources); (3) private industry (through Weyerhaeuser Canada Ltd.); (4) to (6) three aboriginal groups (through the Federation of Saskatchewan Indian Nations, Prince Albert Grand Council and Montreal Lake Indian Band); and (7) a professional forestry group (Canadian Institute of Forestry). Thus it represented a full spectrum of philosophies, cultural backgrounds, social values, management intents, legislation and policies, and economic goals (PAMF Association, 1992, p. 5). The following goals of the PAMF were proposed:

- to ensure a sustainable and predictable supply of forest-based ecological, social and economic benefits through the management of forest ecosystems and people; and
- to raise awareness of and commitment to the concept of sustainability, integrated resource management and ecosystem management among forest users, researchers and managers at the local, regional, national and international levels.

The PAMF partnership also suggested that 'the partnership is committed to accomplishing sustainable development of the forest lands through the application of integrated resource management as a tool … the partners recognize the need to initiate an adaptive[6] and predictive management strategy' (ibid., pp. 20–21).

The management of the PA Model Forest has been by a Partnership Management Committee, with the assistance of a full-time administrator. The Management Committee has equal representation from all seven partners, with the chair of the Committee selected annually.

Land Use in the PAMF Region

The PAMF is located in central Saskatchewan on 309 720 ha of boreal forestland. Its location and vicinity are shown in Figure 8.1. It is located 70 km north of the city of Prince Albert, and is accessible by major highways. Soils are primarily luvisols associated with glacial tills and lacustrine deposits with some brunisols and minor chernozemic soils (PAMF, 1992, p. 8). The hydrology of the region consists of several lakes, which form a part of the Churchill River Basin. The climate of the region is characterized by long, cold-to-very-cold winters, followed by short, warm-to-very-warm summers. Precipitation during the winter months is relatively low. Average precipitation is 320–60 mm, of which 70 per cent is in the form of rain (PAMF, 1992, p. 9).

The PA Model Forest region consists of three separate jurisdictions: (1) the area occupied by the Prince Albert National Park (PANP), under the jurisdiction of the Canadian Park Service; (2) a commercial forested area, under the jurisdiction of Weyerhaeuser Canada Ltd; and (3) wooded and non-wooded (such as lakes) areas of the Montreal Lake Cree Nation (MLCN) Reserve under the jurisdiction of the MLCN Band Council. The commercial forestlands, which

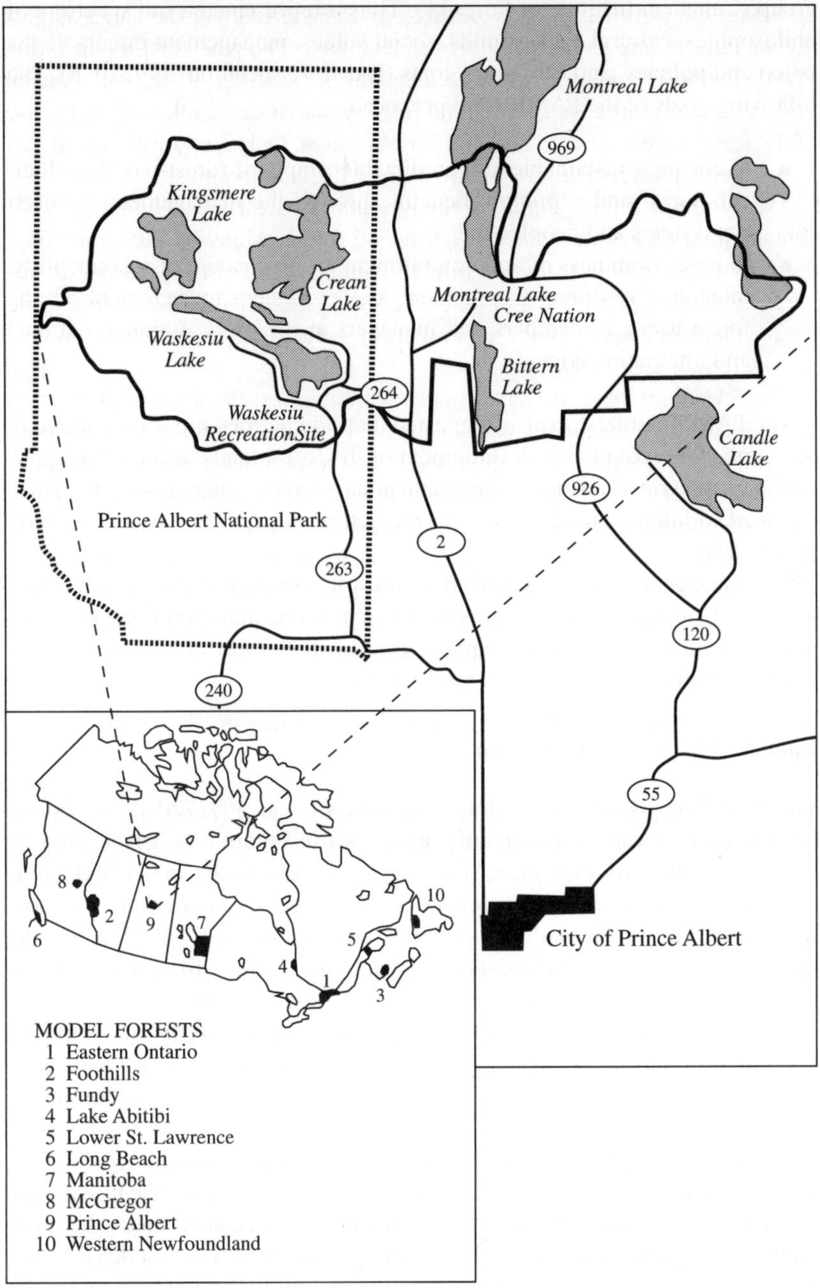

Source: Kulshreshtha, Siemens, Doell and Walker (1994, p. 3).

Figure 8.1 Map of the Prince Albert Model Forest region

are under the jurisdiction of Weyerhaeuser Canada Ltd, are the single largest region, constituting roughly half of the total area. The other major part of the PAMF region is a natural reserve, under the jurisdiction of the PA National Park. Within this area, no commercial forestry activities are allowed, as the use of the forest is restricted to non-consumptive uses. Only a relatively small part of the region is under the jurisdiction of the MLCN authorities.

The nature of timber supply in the productive forests is such that 35.6 per cent of the total growing stock is hardwood forest, while the remaining 64.4 per cent of the total is softwood forests. Among the hardwoods, aspen dominates the landscape in terms of mature growing stock, while jack pine, white spruce and black spruce are the dominating softwood species (based on data provided by WCL, 1989).

The PAMF region houses two major communities: the town of Waskesieu Lake, located within the boundary of the PA National Park, and the community of Montreal Lake and its three satellite communities, Little Red Reserve, Weyakwin and Timber Bay. Of these, the community of Waskesieu Lake is a resort community, where the population swells during the summer months. However, the park administration and associated services support an all-year-round population of 212 people (Kulshreshtha, Moriarty and Walker, 1994, p. 25). The population of the Montreal Lake Cree Nation (MLCN), based on a survey of these communities, was estimated at 1132 people.[7]

Major Economic Activities and Values Associated with the PAMF Region

The PAMF region provides various direct and indirect economic services to local people, as well as to other members of society at large. Many of these services result in an economic value. Various activities related to forests were classified using the 'Total Economic Value' approach, which include use-related values and non-use-related values, plus various intrinsic values[8] of the forests. Many of these values are related to the ecosystem, and tend to be more global in nature.

Concentrating on the use-related values, forests in the PAMF generate two types of activities: production of market goods and generation of non-market goods. Various timber and non-timber goods are included among the market goods. Non-market goods include those products whose value (price) is not determined by market forces. These products include recreation at the PA National Park, hunting in areas outside the PANP, bird watching, and food gathering and traditional uses of the forests by the First Nation people. In addition, some wilderness-based recreation activity and ecotourism is also present in the region, although the magnitude of such activities is not well documented.

Economic activity related to timber products

Timber products produced within the PAMF include logging, collection of fuelwood, processing of timber into lumber and pulp, and production of

Table 8.1 Level of economic activity and value of forest in production of timber products, 1993

Forest use	No. of permit holders/ users	Level of producer surplus in dollars: Per operation	Level of producer surplus in dollars: Per m^3 of timber
Timber Products			
Logging	271	32 742	20.56
Commercial fuelwood production	23[a]	505	0.22
Non-commercial fuelwood production	20[a]	–607	–4.58
Pulp production[b]	1	–26.22[g]	–5.40
Paper production[b]	1[c]	–78.30[g]	–16.11
Sawmilling[b]	1	17.04[h]	7.22
Other wood industries	0[d]	—	—
Non-Timber Products			
Trapping	11	888	—
Outfitting	55[e]	2 599	—
Speciality forest products	10[f]	700[i]	—
Grazing	58	2 312	—
Wild rice production	25[f]	2 167	—
Commercial fishing	50[f]	445	—

Notes:

— Not applicable or not available.

[a] Data on actual distribution of commercial and non-commercial fuelwood permit holders not available. The sample distribution of 45 per cent of all fuelwood permit holders being non-commercial was used, while the rest were treated as commercial operations.

[b] Estimates are for the year 1991.

[c] Single integrated operation of pulp and paper production.

[d] None located within the PAMF region; however, several were located in the surrounding region.

[e] Includes outfitters who are located in the surrounding region.

[f] Number somewhat uncertain, information on this sector is very poor.

[g] Values are dollars per tonne.

[h] Values are dollars per thousand boardfeet.

[i] Estimate based on a very small sample, thus not representative.

Source: Kulshreshtha (1995a).

paper. In addition to these, several other wood processors are located in the surrounding region, but not directly within the PAMF. Details on the importance of these activities are shown in Table 8.1.

Much of the commercial activity is related to the production of lumber, pulp and paper. The Big River Sawmill and the Prince Albert Pulp and Paper Mill are the two major wood processors for the region, although they are not located within the physical boundary of the PAMF region.

Non-market goods within the PAMF

The PAMF region, on account of the PA National Park, provides a variety of non-market goods. Various studies on estimating a value of recreation and wilderness preservation for the region have been carried out. For example, Loewen and Kulshreshtha (1995a) estimated the value of recreation activities to be between \$13.68 and \$24 per person per user-day. In addition, a non-use value of \$15.3 million for wilderness preservation was also estimated (Loewen and Kulshreshtha, 1995b). A distinct feature of this valuation was the divergence between the valuation of aboriginal and non-aboriginal families with respect to wilderness preservation. The aboriginal families were prepared to pay, on average, \$81 per family for current preservation, about \$20 more than non-aboriginal families (Loewen and Kulshreshtha, 1995c). Although other non-market goods, such as the existence and bequest values, are important, no credible estimates of such values are available.

MAJOR THRUST OF THE PAMF ACTIVITY

Financial Resources and their Allocation

As noted above, the establishment of the Model Forest Program was based on a financial commitment by the federal government. The PAMF Association's activities were financed through such a grant, as well as through other cash and in-kind contributions. During 1994/5, the total assistance from the federal government was \$1.16 million, of which almost two-thirds (63 per cent) was spent on research and development, another 23 per cent on communications, and the rest on administration.

In addition to the above contributions, the PAMF also received \$85 000 worth of funds from other federal and provincial agencies, and \$73 000 through collaborative research and other sources, for a total contribution of \$1.325 million. In addition, the PAMF reported a total of \$0.989 million as in-kind contribution.

Focus of Research Activities

The PAMF research activity concentrated on three fronts: (1) inventory of resources, (2) understanding the ecosystem, and (3) socioeconomic information. The first set of activity included development of a Geographical Information System, which was used to estimate the current status of forest resources, and later on to prepare maps of the region, highlighting timber resources. The second set of activities included ecosystem research, biodiversity research and silvicultural research. Ecosystem research included modelling forest productivity, identifying soil and landscape factors affecting forest productivity, and quantification of hydrological pathways, whereas the biodiversity research included genetic diversity, fungal biodiversity and effect of disturbances on plant diversity. In addition, research projects also included study of bird habitats (such as the barred owl as an indicator). Silvicultural research included site preparation methods and modelling wildlife habitats.

The socioeconomic research for the PAMF was carried out on three major themes: cultural values and history of the Montreal Lake First Nations people; economic valuation of forest resources and uses, including prediction methods; and economic opportunities identification for the First Nations people in the region. The research was undertaken in this particular sequence.

Distinguishing Features of the PAMF

The PAMF is incorporated as a non-profit organization. The overall land use decisions on provincial Crown lands remain the responsibility of the provincial government, and, on the federal Crown lands, of the PA National Park. This excludes the reserve lands given to the First Nations people, which are under the jurisdiction of the MLCN Band Council. On provincial Crown lands, the day-to-day execution of forest management activities, however, is carried out by Weyerhaeuser Canada under a Forest Management License Agreement (FMLA). There are some distinguishing features of the PAMF, which are shown in Table 8.2. The focus of research and other activities of the Association support these features. Conceptually, the PAMF adopted the management approach based on knowledge, and, where such knowledge was a constraint, resources were channelled into research and development. Although some activities were undertaken to disseminate the PAMF mandate to the public at large, much of the participation at the model forest level was among the seven partners. Similarly, various types of values – economic, social, cultural, intrinsic – were recognized, perhaps because of the unique setting in which the partnership had to operate. A case in point may be the recognition of the non-market goods and value of wilderness preservation by

Table 8.2 Distinguishable features of the Prince Albert Model Forest

Particulars	Model Forest Details
Ecosystem management approach	Based on scientific understanding of natural ecosystems; incorporates traditional aboriginal land ethic and protection of national park.
Public participation	Largely through representative organizations; communication tools used to increase public interest.
Knowledge orientation	Basic ecosystem knowledge; socioeconomic studies to improve equity and promote healthy communities; traditional and scientific knowledge are valued equally.
Integration of values	Resolution of competing specific values held by national park, industry and aboriginal community has been primary focus; park addresses non-market values related to biodiversity.

Source: Brand *et al.* (1996, p. 84).

the PA National Park, which happens to be a partner, occupying almost half of the total PAMF region.

Selected Accomplishments of the PAMF

As a direct or indirect result of the formation of the PAMF Association, where the seven partners came together to discuss their issues and problems, several events have happened which are worthy of noting.[9] The first one of these is the creation of the Aboriginal Employment Equity Program by the Prince Albert National Park. Under this programme, workers of First Nation ancestry are hired for interpretation services. The second noteworthy change involves Montreal Lake Cree Nation members and Weyerhaeuser Canada. They have developed at least two cooperating activities: (1) planning of harvest in a region adjacent to the community was carried out jointly by the Weyerhaeuser team and local people,[10] and (2) awarding of contracts to reserve-based companies for silviculture. In addition, the PAMF has also initiated training of young students from the MLCN in a GIS (Geographical Information System) literacy programme, which may have enhanced their forestry-related technical skills, and prepared them better for the job market.

EVALUATION OF THE PAMF: MAJOR ISSUES

Before proceeding to the evaluation of the PAMF, one should keep in mind the limited amount of time that has elapsed between the initiation of the PAMF and its evaluation at this time. The Model Forest Program has only been in existence for four years. Given that the PAMF did not get approved until April 1993 and, furthermore, that a significant amount of time is consumed by initial planning and preparation activities, this leaves at most a three-year time period for achieving the PAMF objectives. Information on the last year (1996/7) is scanty at this time. Thus this evaluation is based on two complete years (from 1994/5 to 1995/6) of operation – a rather short period of time to assess long-term implications of such a programme.

Evaluation Issues

Evaluation of the PAMF can be carried out using several criteria. However, the most relevant criterion is that of the programme relevance – whether the PAMF has achieved the Canadian Model Forest Program objectives, as well as its own stated objectives. In this study, in addition to programme relevance, three other criteria were added: programme effectiveness, programme management and programme impacts. For each of these criteria, a number of questions were selected, which are shown in Table 8.3. On account of unavailability of data, discussion of these questions is limited to observations based on personal knowledge of the author or based on some 'expert opinions'. The limitations of these sources of information must be kept in mind in interpreting the conclusions reached.

Programme Relevance

As noted earlier, Canada's Model Forest Program was designed to accelerate the implementation of sustainable development in the practice of forestry. In light of the PAMF's stated objectives of commitment to sustainability, and integrated resource management at various levels, the goals of the PAMF appear to be consistent with those of the Model Forest Program. The federal government also initiated another programme under the Green Plan, the Decision Support System (DSS) programme. The DSS programme was designed to develop an innovative set of DSS models to meet the needs of Canadian forest managers (Wedeles and Kurz, 1992, p. 1). Although the DSS Group developed a number of models, their actual application in solving issues facing Canada's model forests, including the PAMF, remains to be seen. A partial explanation for this lack of integration may be the process that was followed in developing the DSS models. The DSS Group planned such models on the basis of their

Table 8.3 Selected evaluation issues for the PAMF

Category	Relevant question(s)
Programme relevance	To what extent were the objectives of the PAMF consistent with those of the model forest programme?
	How did the PAMF activities relate to other forestry-related initiatives of the federal/provincial governments?
	Did the PAMF succeed in adopting sustainable development and/or developing integrated forest management?
Programme effectiveness	Has the PAMF been successful in achieving the stated objectives?
	Were various activities of the PAMF consistent with achieving the objectives of the PAMF?
	Did the PAMF succeed in creating an effective partnership?
	Has the PAMF improved the planning capability for forest resources?
	Did the PAMF improve the public perception of forest management and Canada's Model Forest Program?
	Were the research activities consistent with the concept of integrated resource management?
Programme management	How effective was the overall management of the model forest?
	How effective were the management of research and other activities of the PAMF?
Programme impacts	What impacts, intended or unintended, resulted from carrying out the activities of the PAMF?
	Has the programme resulted in networking among various agencies for sustainable forest management?
	Has the PAMF resulted in changes in the legislation or provincial policies regarding forest management?
	Did the PAMF activities overlap with other provincial and local programme activities already in place?

own merits, anticipating their use by the model forests, rather than planning them after carrying out a need assessment of the model forests (including PAMF) with respect to applicability of such models.[11]

With respect to the question of success of the PAMF in developing integrated forest management, and adopting sustainable development in the management of forests, the answer appears to be that it has been rather limited. It can be stated that the PAMF partners perhaps are familiar with these phrases and, in principle, do not reject them in the context of forest management in the region. However, as of this time, there is little evidence to suggest that the PAMF partners have reached an agreement as to what constitutes sustainable forest management in the PAMF region, and how to implement it within the PAMF region.

Programme Effectiveness

The goals of the PAMF Association were translated into 23 working objectives. According to the PAMF Association (1995), various activities that were undertaken were related to these objectives. To what extent the completed activities have met the objective(s) cannot be answered in a definitive manner at this time, on account of limited available information. However, is should be safe to state that the Association has directed its research and communication activity on lines suggested by these objectives, and some limited success in meeting these working objectives has been achieved.

The PAMF did succeed in creating a multi-stakeholder process through the creation of the partnership. In fact, this partnership was formed in the very beginning and has continued until this time. As to the question, 'Is the partnership an effective one?', evidence is unclear. During the course of planning PAMF research activities, there were instances when one of the partners did not permit use of resources under their jurisdiction for the PAMF-approved research project, in spite of the fact that the project was approved initially by various committees of the PAMF in which the very same partner institution had a representation. In most cases, instances of a partner's cooperation in PAMF research activities were limited to those where the partner sponsored that research activity directly and/or had a vested interest. In fact, there were research activities in which one of the partners was interested for private[12] reasons. This type of activity included cultural and natural history of the Montreal Lake Cree Nation, which was undertaken because of the perceived need for educating the students of the MLCN in these areas. A very similar example was that of an economic opportunities identification project which was sponsored by the MLCN as well, and sought opportunities for local workers in avenues which were not necessarily related to forest resources of the region.

There is evidence to suggest that the general public in close proximity to the city of Prince Albert, and those associated with some aspects of the forest industry, are relatively more aware of the Model Forest Program now than at the time of its inception. The PAMF has planned several activities to inform the public of its mandate and research activities. However, it is the opinion of this author that, in general, the public perception of the PAMF remains poor. This was demonstrated during the course of various surveys that were undertaken to estimate the economic value of alternative forest uses in 1993–4. A majority (over 75 per cent of the respondents) of the local forest users – grazing permit holders, trappers and wild rice producers, among others – were unfamiliar with the PAMF: its role and purpose. Even among those who are more closely dependent on the forest resources of the region – the primary and secondary wood processors (excluding Weyerhaeuser Canada) – a common view was that the PAMF is a government body and, therefore, the information being solicited will be used against them in future.

One of the important aspects of the Model Forest Program was that of integrated resource management. One of the prerequisites for such a development is establishment of an appropriate knowledge base. Did the PAMF attempt to establish such a knowledge base? The answer to this question appears to be affirmative, but collected information was non-integrated. A number of reasons can be cited for this. First, many of the research projects were not exclusively sponsored by the PAMF. These were ongoing or institutionally approved activities, and the PAMF sites were added to these projects as a direct result of financial assistance from the PAMF. Second, most of the projects for ecology or socioeconomics tended to be single objective, with no obvious integration possibilities. In some cases, projects were of interest to only one of the partners of the PAMF. For example, projects dealing with the wood supply, ecosystem and other physical and biological aspects did not include socioeconomic aspects. Similarly, the studies sponsored by the MLCN dealt exclusively with socioeconomic aspects, with no interrelationships with natural and physical aspects of the forests or other regional natural resources. Third, the PAMF established an Integration Committee during the fall of 1995. The Committee met to review the possibility of integrating various aspects of research completed or currently in progress. After due deliberations, the Committee concluded that such integration was not possible, since there was no common base to work with.

Programme Management

The overall management of the PAMF was the responsibility of the Board of Directors, which was assisted by a seven-member partnership Technical Committee. A four-member Research Committee, which came into existence during the summer of 1994, assisted the Technical Committee as well.[13] Most contractual arrangements for the projects were made between the project investigator(s) and the Technical Committee. The role of the Research Committee was to assess the professional contributions of completed or in-progress projects. However, the evaluation of the Research Committee suffered from two problems. First, the Research Committee had limited inputs into the research programme of the PAMF, especially during the first two years. This perhaps steered the activities in a direction that may not have been the Research Committee members' first choice. Second, in some cases, the criticisms of the Research Committee tended to be idealistic, without any regard to the funding available for completion of the task at hand. For example, some of the reports in socioeconomics were criticized on account of the poor response rate. However, for mail-in surveys (which was the only method possible within the budget provided) in northern communities, such responses are typical. Three, the Research Committee also did not adequately take into account the lack of cooperation by member partners in completing the PAMF approved research projects.

Programme Impacts

Among various aspects of evaluation of the PAMF, this aspect is a more difficult one, on account of the shorter time period. In spite of this, some examples of unexpected impacts of the PAMF can also be identified, which include, but are not necessarily limited to, the following.

- Development of partnership for consultation in other matters, besides forest management issues.
- Opening of research and development opportunities on forest resources for academic institutions in the province.
- Richer education of First Nations school children in their cultural heritage.
- Organization of an International Conference during the summer of 1995, with some economic spin-offs to the city of Saskatoon, as well as providing avenues for research activities related to forests through new or renewed contacts.

The expected impacts on improved forest management will undoubtedly develop out of some of the activities that are already in place, such as the aboriginal equity programme, and joint planning of forest harvesting, besides generating additional economic activities on the MLCN reserve. At this time, such impacts are hard to identify.

Although the PAMF has resulted in a working partnership, whether that has further evolved into a cohesive network of agencies for carrying out sustainable forest management is not clear. Various partners[14] also have related activities that are not included in the mandate of the PAMF. A situation of overlap could potentially exist, unless steps are taken to minimize such possibilities. In terms of impacts on legal aspects, particularly on provincial or regional legislation, the limited available evidence does not support such impacts.

LESSONS LEARNT

The Prince Albert Model Forest, like other model forests in Canada, is a working laboratory for developing integrated resource management and sustainable forest management. Thus it has the potential to provide guidance in terms of its evaluation – the strengths and weaknesses of the approach. However, one must be cautioned against the small (only one) sample, and a short observation period (only two years) on which these conclusions are based.[15]

The Partnership Approach

The Model Forest Program has created a unique partnership for managing the forest resources. In the case of the PAMF, this included not only the two levels of government (which are traditionally involved) and the industry, but also the First Nations people. This partnership has brought different cultural and economic values and objective functions. In spite of these differences, the partnership has survived since its inception, which is strong positive evidence of the success of this approach. Many of the accomplishments noted above indicate that such institutions can work. Members of the partnership are willing to listen to other stakeholders and partners, their points of view and vested interests in the partnership, and have tried to resolve some of them.

Effectiveness of an Extension Partnership Approach for Forest Management

Although the PAMF partnership has worked for a smaller region, the obvious question is whether it would be equally effective if the approach was extended to the entire forest region. An answer to this question cannot be given with any degree of accuracy. However, theoretically, one can argue that some members of the partnership may be more willing to follow a consensus approach (such as that followed in the case of PAMF) for a smaller region. Whether a consensus can easily be reached for a wider region remains to be seen. More noteworthy is the role of the PANP-type activities related to nature preservation. Since the PANP constitutes half of the total area of the PAMF, such activities are prominent. Whether the very same degree of prominence could be expected if the model forest regions were to include the whole of northern Saskatchewan is questionable.

Sustainable Forest Management and Partnership Approach

Generally speaking, a group consensus is less likely to be reached on complex concepts, such as sustainable forest management. This is because of two problems that are encountered in this respect: (1) it may not be reasonable to expect a group of stakeholders to define such concepts in a clear and concise manner; and (2), even if it is defined, it may be difficult for a group of stakeholders to agree to it. Furthermore, according to Carpenter (1993), the concept of sustainable forest management is even more difficult to measure programmatically. According to Eckman (1993, p. 277), unsustainability is easier to identify, measure and address at the project level. Perhaps the model forests could attempt to develop unsustainability guidelines initially.

Development of sustainable forest management demands some knowledge of long-run consequences of adopting alternative management practices. For the PAMF region, such a knowledge base is poor, although it is improving as time progresses. For the PAMF, Bouman *et al.* (1996, p. 63) examined the existing knowledge base and concluded that 'The landbase, productivity and forest value approaches are proposed to advance sustainable forest use; but current data availability favors the landbase approach.'

Integrated Forest Management and the Group Consensus Approach

Development of integrated forest management through a group consensus approach suffers from limitations similar to the development of sustainable forest management. Integration of resource management requires understanding multiple (social, economic and ecological) values of the resources in alternative uses, and developing a mechanism for their incorporation in the decision-making process. This requires measurement of trade-off levels that result with the resource being used in a specific manner, and a weighting scheme to present the final outcome in a common unit. The experience of the PAMF indicates that such a trade-off analysis is lacking, which may impede the process of integrated resource management.

Lack of a Mechanism to Endogenize all Values of the Forest

As alluded to previously, forests have many values; some are related to economic activities, while others are more intrinsic in nature. A basic question is, 'Whose values should be counted in the decision-making process?' For sustainable development, values of both current and future generations should be included. The question is how.

Even if we could solve the issue of whose values, the question of how society convinces those in power to consider the values of people who are not participating in the decision making is very relevant. Typically, such values are introduced through regulatory practices, but our experience has shown that regulations are not the most cost-effective way of ensuring mutual respect among various members of the society. Furthermore, this raises the jurisdictional issue. Some of the values of the forest are global in nature (such as the value of forest for carbon sequestration). How does the world community convince those who are doing the damage in this regard?

Legislative Changes should Coincide with, if not Precede, Needed Change

Integrated resource management is often hindered by the existence of provincial and federal regulations. One such example in forest harvesting is that, in order to maintain some wildlife in the harvested forest regions during the post-harvest period, somewhat taller tree stumps could be left. However, this violates the provincial policy where a tree stump higher than three inches is illegal. Thus it is imperative that more emphasis be placed on identifying current legislative bottlenecks to integrated resource management and sustainable forest management.

Property Rights

Related to the legislative aspects of forest management is the issue of mandate to make management/operational decisions. In the case of the PAMF region, under the present system, the private company is awarded a forest management licence, which gives it the property right to decide and choose the use of the forest which best meets its objective. Although the award of the licence is with some conditions, these are not set in consultation with other stakeholders in the model forest region. Private industry decision making is generally driven by its responsibility to the owners of the company: shareholders. The model forest philosophy/concept would have a better chance of success in adopting integrated resource management if such an integration plan preceded the award of the licence. This modified system is shown in Figure 8.2. If the licence is awarded after the institutions such as the model forest have developed the integrated resource management guidelines, the private company will be obligated to adhere to them, if mandated under the terms and conditions of the licence.

Conflict between Group and Private Information in Decision Making

In many partnerships, their ability to work together depends upon the degree of trust that has developed in making group decisions. Provision of appropri-

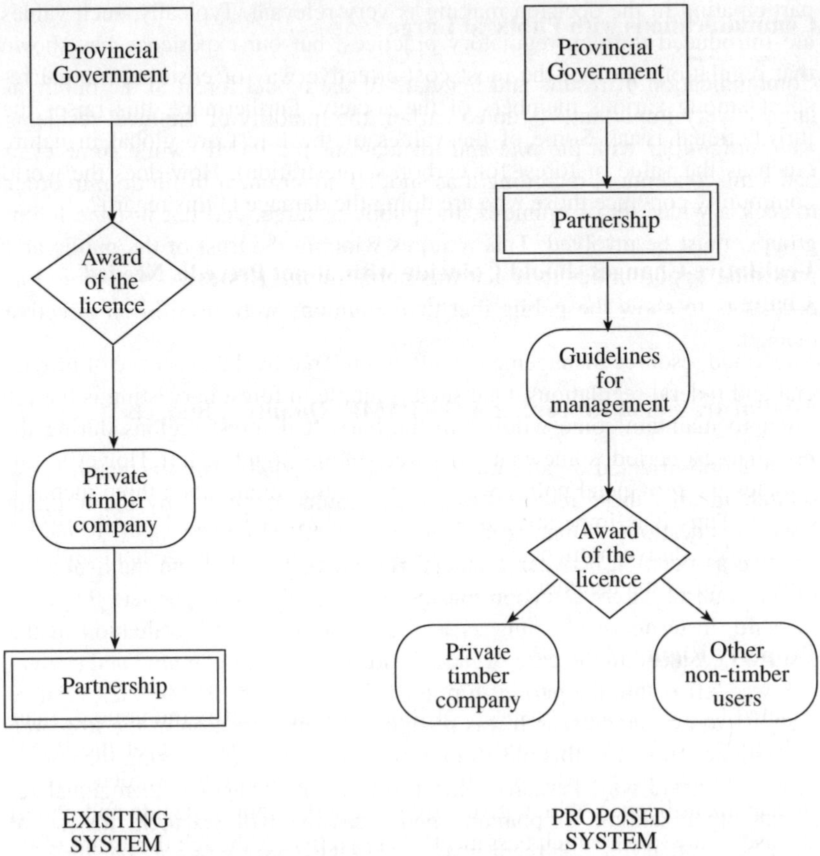

Figure 8.2 Alternative forest management philosophies

ate information is a key to this trust. If some partners do not share that information, working of the partnership will simply become a place for debates, and not a mechanism for effective decision making. In the case of the PAMF, given that some such problems exist (single stakeholder sponsorship of research projects, inability of some researchers to get adequate cooperation from other partners, among others), it appears that such a common understanding for the PAMF as a whole may not have been achieved. In any group situation, conflicts are natural. However, groups must also devise means of conflict minimization, which requires some efforts in conflict management.

Communications with Public at Large

Communication of results and mandate of the model forest to the public at large is very important. As noted earlier, the majority of the local residents were unfamiliar with the role and mandate of the PAMF, while some even had a misconception, regarding it as another government institution. In order to seek a wider set of opinions, the public at large, and not just the lobby groups, must be involved. This requires winning the trust of the public and providing opportunities to solicit this public input. However, even more important is to show the public that their opinions were used in an effective manner.

Administration of Research for the PAMF: Quality v. Budgets

In an action-oriented organization, provision of good-quality research is essential. Ideally, these research activities should be guided by the ultimate issue(s) (question(s)) that need(s) to be solved. However, there is also a positive association between financial resources provided and the quality of the ensuing research. Decision makers must be able to appreciate this relationship, in terms of planning as well as monitoring and evaluation of the research projects. In the case of model forests, research planning and evaluation was left within the partnership itself. In the case of PAMF, their relative inability to evaluate research was recognized in the creation of the Research Committee. However, this meant loss of continuity, and changing the course of research mid-way. Perhaps some institutional linkages with regional research organizations for planning and assessment of research, instead of carrying out that function themselves, might have been mutually beneficial.

The above list of lessons learnt is not comprehensive and many other items can be added to this list, as more data and evidence are gathered.

SUMMARY AND CONCLUSIONS

The Prince Albert Model Forest was established under the Canadian Model Forest Program, which was initiated on account of several concerns in the public arena with the way forests were being managed currently. The PAMF is a partnership of seven stakeholders, representing various levels of the government, industry and the First Nations people, in addition to a professional body. The area under the jurisdiction of the PAMF is small, only 315 326 ha. In spite of the small size, the forest provides both timber and non-timber products. The goals of the PAMF were, first, to ensure a sustainable and predictable supply of forest-based benefits (ecological, social and economic) through the management of forest ecosystems and people; and, second, to raise awareness of the concept of sustainability, integrated resource management and ecosystem management among forest users, researchers and managers of the forests.

The PAMF undertook a significant amount of research on three fronts: inventory of resources, understanding ecosystems, and socioeconomic information. In addition, communication activities were planned to raise awareness of the model forest activities and its mandate.

The success of the model forest concept can be viewed in terms of the continuation of a working partnership among the seven stakeholders. The very fact that the original seven members decided to continue with the partnership, and added three new members in Phase 2 of the project, is solid enough evidence supporting such a claim. In addition, during this phase, several cooperating activities have been initiated. The Model Forest does act as a forum to bring forth concerns of other stakeholders. However, it cannot be said with any degree of certainty whether the philosophy of forest management would change if the entire Forestry Management License Area (or the entire northern part of the province) was to be managed on these lines. Sustainable and integrated forest management also requires good information. Although significant strides have been made in collecting the missing information, significant gaps persist. Many of these gaps are in terms of applying the scientific knowledge to solve various issues leading to sustainable forest management in the region.

NOTES

1. Original paper presented at the Costa Rica Workshop of the IAI Initiative, CATIE, Turrialba, 27 February–1 March 1997.
2. This is because each of the model forests is situated in different types of forest ecosystems, and a diverse mix of stakeholders.
3. Based on Natural Resource Canada (1996). For more details on the programme, see this report.
4. There appears to be some confusion as to the exact status of these model forests. The IMFN (1996) indicates that these are in the process of developing, whereas Natural Resource Canada suggests that these actually joined the network in 1995.
5. Over a period of time, the composition of the partnership changed. The Lac La Ronge Indian Band was a signatory for the proposal, but did not continue after the approval of the proposal by the federal government.
6. Adaptive management ensures that the forest management activities, including cause and effect relationships on ecological, economic ad social benefits, are monitored and compared against the desired benefits.
7. This is based on results of a survey reported in Kulshreshtha (1995b). The average family size at the Montreal Lake reserve was 4.58 people, whereas at the Little Red River, Weyakwin and Timber Bay it was estimated to be 4.18, 3.65 and 3, respectively. Using the estimated number of families, based on the information provided by the Band Council, total population of the MLCN was estimated at 1132 people.

8. The intrinsic values, according to Rolston (1983), refer to those that are not assigned by human beings, but are inherent in the object or its relationship to other objects. These values include various values in terms of maintaining/changing the ecosystem. Some of these values have relationships with the human and social system, and can be recognized as economic values.
9. This listing excludes cooperating activities between the seven partners and other organizations or research bodies that did not go through the PAMF approval process.
10. For more details on this planning activity, see Weyerhaeuser Canada (1997).
11. Similar statements can be made for other model forests in Canada. Many of the models developed were complex and their translation into application by the model forest personnel was not an activity given a high priority by the DSS group.
12. The term 'private' here is used to differentiate a partner's objective from the PAMF Association's objective, as well as from those in which more than one partner had an interest.
13. It should be noted that by this time research activities for the first two years of operations were already finalized by the Technical Committee.
14. This will include various forest harvesting and silviculture activities of the Weyerhaeuser Canada, and nature preservation activities with the PANP.
15. Furthermore, it should be emphasized that these remarks are not aimed exclusively at the PAMF, but perhaps at the Model Forest Program as a whole, whether in Canada or internationally.

REFERENCES

Bouman, O.T., G. Langen and C.E. Bouman (1996), 'Sustainable use of the boreal Prince Albert Model Forest in Saskatchewan', *The Forestry Chronicle*, 72(1), 63–72.

Brand, D.G., O.T. Bouman, L. Bouthiller, W. Kessler and L. Lapierre (1996), 'The Model Forest Concept: A model for future forest management', *Environmental Reviews*, 4(1), 65–90.

Carpenter, Richard A. (1993), 'Can Sustainability be Measured?', *Environmental Strategy*, 5 (February), 13–16.

CCFM (Canadian Council of Forest Ministers) (1992), *Sustainable Forests – A Canadian Commitment*, Ottawa: Canadian Council of Forest Ministers.

Eckman, Karlyn (1993), 'Using indicators of unsustainability in development programs', *Impact Assessment*, 11(3), 275–88.

IMFN (International Model Forest Network) (1996), *IMFN Newsletter*, Ottawa, Ontario.

Kulshreshtha, S.N. (1995a), *Economics of Alternative Forest Uses: A Case Study of the Prince Albert Model Forest Region*, Prince Albert: The Prince Albert Model Forest Association.

Kulshreshtha, S.N. (1995b), *Socio-Economic Baseline Survey of Montreal Lake Cree Nation*, Prince Albert: The Prince Albert Model Forest Association.

Kulshreshtha, S.N., B.F. Moriarty and H.V. Walker (1994), *Considerations Involved in Developing an Economic Assessment Framework for the Prince Albert Model Forest Region*, Prince Albert: The Prince Albert Model Forest Association.

Kulshreshtha, S.N., J.K. Siemens, W.S.M. Doell, and H.V. Walker (1994), *Economic Perspective on the Prince Albert Model Forest Region of Saskatchewan*, Prince Albert: The Prince Albert Model Forest Association.

Loewen, K.G. and S.N. Kulshreshtha (1995a), *Economic Value of Recreation Experi-*

ence at the Prince Albert National Park, Prince Albert: The Prince Albert Model Forest Association.

Loewen, K.G. and S.N. Kulshreshtha (1995b), *Recreation and Wilderness Participation and Economic Significance in Saskatchewan*, Prince Albert: The Prince Albert Model Forest Association.

Loewen, K.G. and S.N. Kulshreshtha (1995c), *Economic Aspects of Wilderness Valuation and Recreation Uses by Aboriginal Households: A Case Study of Prince Albert*, Prince Albert: The Prince Albert Model Forest Association.

Natural Resources Canada (1996), *Model Forest Network: Year in Review 1994–95*, Ottawa.

Nault, Robert (chair) (1994), *Canada: A model forest in the making*, Report of the Standing Committee on Natural Resources, House of Commons, Canada, Issue no. 25.

PAMF Association (1992), *Prince Albert Model Forest*, Prince Albert, February.

PAMF Association (1995), *Third Annual Report of the Prince Albert Model Forest Association Inc.*, Prince Albert, Saskatchewan.

Rolston, H. (1983), 'Values Gone Wild', *Inquiry*, 16(1), 181–207.

WCED (The World Commission on Environment and Development) (1987), *Our Common Future*, New York: Oxford University Press.

WCL (Weyerhaeuser Canada Ltd) (1989), *Twenty-Year Management Plan: 1990–2010*, Prince Albert.

Wedeles, C. and W. Kurz (1992), *Preliminary Integration of Forestry Canada's Model Forest Program and Decision Support System Program*, report prepared for Forestry Canada, Richmond Hill, Ont.: ESSA Environmental and Social Systems Analysts Ltd.

Weyerhaeuser Canada Ltd (1997), 'Model Partners', *Saskspirit*, 9(1), 8–9.

APPENDIX: SELECTED CHARACTERISTICS OF THE CANADIAN AND INTERNATIONAL MODEL FORESTS, 1996

No.	Name of the model forest	Date of joining	Area in thousand ha.	Forest region	Province/state and country
Canadian Model Forests					
1	Western Newfoundland Model Forest	May 1993	707	Boreal	Newfoundland, Canada
2	Fundy Model Forest	April 1993	419	Acadian	New Brunswick, Canada
3	Lower St. Lawrence Model Forest	May 1993	113	Great Lakes – St. Lawrence	Quebec, Canada
4	Eastern Ontario Model Forest	February 1993	1 534	Great Lakes – St. Lawrence	Ontario, Canada
5	Lake Abitibi Model Forest	June 1993	1 095	Boreal	Ontario, Canada
6	Manitoba Model Forest	June 1993	1 047	Boreal	Manitoba, Canada
7	Prince Albert Model Forest	April 1993	315	Boreal	Saskatchewan, Canada
8	Foothills Model Forest	March 1993	2 500	Boreal, Montane, Subalpine	Alberta, Canada
9	McGregor Model Forest	April 1993	181	Montane, Subalpine	British Columbia, Canada
10	Long Beach Model Forest	September 1994	400	Coast	British Columbia, Canada
International Model Forests					
11	Chihuahua Model Forest	August 1993	110	Temperate	Chihuahua, Mexico
12	Calakmul Model Forest	August 1993	380	Tropical	Campeche, Mexico
13	Mariposa Monarca Model Forest	February 1995	477	Temperate	Michocan, Mexico
14	Gassinski Model Forest	October 1994	384	Temperate	Khabarovsk, Russia
15	Bentong Model Forest	June 1995	183	Tropical	Pahang, Malaysia
16	Cispus Model Forest	April 1995	60	Coastal temperate	Washington, USA
17	Applegate Model Forest	April 1995	115	Coastal temperate	Oregon, USA
18	Hayfork Model Forest	April 1995	203	Coastal temperate	California, USA

Source: Natural Resource Canada (1996, pp. 24–5).

9. Forest policy in Costa Rica[1]

Olman Segura-Bonilla

INTRODUCTION

In Costa Rica, as in other Central American countries, the forest has been irrationally exploited and, if the current style of management is not changed, the maintenance and recuperation of forest cover will be increasingly difficult. According to several studies, as we will see below, it would be ideal if, in the future, Costa Rica could maintain and even expand its natural systems by means of policies that are clearly understood. In order to reach this goal, profound institutional and economic changes must be made in the society. The institutional changes refer to customs, routines, behaviour patterns and the way resources are used. The economic changes correspond to the creation of new systems of incentives and alliances for the forest sector as well as for the activities that have traditionally competed with the forest.

The process of economic development establishes a legal and institutional environment which directs and influences land use, including forests.[2] Several authors (such as Pasos, 1994; Segura, 1996) have pointed out that in Costa Rica the predominant search for economic growth, and particularly the increase in the export of farm products, has generated a pattern of agricultural development characterized by a strong dependence on energy input and on land which is not suitable for farming. Agricultural expansion into forest land and the use of inadequate production practices usually has caused the overexploitation of this type of land. Forest land was not valued adequately and the value of forest resources was reduced to only the value of the harvestable timber. In addition, since timber was abundant, no improvements were made in practices to increase efficiency and productivity of industrial activities related to this raw material, and even now antiquated practices exist which produce a tremendous amount of waste.

The legal framework for this process of development has important defects that in turn cause problems in institutional operations. For example, environmental goods and services coming from the forest are not reflected in market values and are therefore not valued; land use practices are not ecologically sound, and the environmental impact of human activities is ignored. There appears to be no understanding of the interrelationship of social welfare, equality and ecological feasibility; the current practices are inconsistent with sustainable development as positive and negative externalities of economic activities are not reflected in market prices. As argued by Segura *et al.* (1996), forest resources are considered for timber production only; however, forests also produce other goods and services that permit a better standard of living for rural communities and society in general. For this same reason, it is suggested that the forest sector should be analysed, taking into account its economic and financial contribution to the national economy, and more comprehensively including the socioeconomic and environmental benefits that these resources provide for human beings.

This review of forest policy in Costa Rica presents a comparative summary of the literature on the Costa Rican forest sector. The objective is to update and compare the literature with the purpose of concluding with recommendations for further research. The second section contains a description of the forest resources of Costa Rica, together with some indicators of the aggregate value of lumber in the Costa Rican lumber industry. The third and fourth sections address silviculture and the timber industry, respectively. Incentives for reforestation are analysed in the fifth section. The social actors and the institutions participating in the forest sector are described in the sixth section; in the final section, some conclusions are presented, emphasizing the main restrictions and potential for the development of new research projects within the sector.[3]

THE FOREST SECTOR IN COSTA RICA

Definition of the Forest Sector

The forest sector as such is not defined in the national accounts of Costa Rica. This leads to confusion, since usually, when the operation of the forest sector is discussed or analysed, people refer to different aspects of timber felling and production. This is not unique to Costa Rica. In the case of the forest sector, the national accounts only record timber products and certain non-timber products; the remainder of the non-timber products and environmental services are not reflected in market prices and are therefore not in the accounts, although they are recognized and valued ethically by society.

However, partly due to the Earth Summit (UNCED, 1992), there is a growing awareness of this sort of deficiency. Segura *et al.* (1996) mention some of the *non-timber products and environmental forest services*, which have not been taken into account, mainly due to the lack of market value and lack of knowledge concerning the contributions being made in Costa Rica. In the case of the *non-timber products*, for example, in economic policy design, many of the wildlife species which are used for food, and other non-timber products used for industry, construction and handicrafts were not taken into account. But that has never prevented indigenous communities and other cultural groups from taking advantage of these products. Similarly, no market value is placed on several medicinal products such as *psychotria emetica (cephaelis emetica* or *c. ipecacuana)*, Chinese root (*smilax spp.)*, camomile and sarsaparilla. The same thing has happened with some construction materials such as vines, palms, lianas and other materials which have traditionally been used by indigenous and rural communities, and are now demanded in large quantities by the tourist industry for the construction of huts and bungalows. Ornamental plants are also important. Although they are cultivated in nurseries and thus given a market value, the seeds are obtained from tropical forests, and yet no commercial value is placed on them.

Segura *et al.* also mention some of the forest's *environmental services* such as the capture of carbon dioxide, the preservation of biodiversity, the prevention of soil erosion, scenic beauty that promotes ecotourism, the maintenance of the water cycle and other services. As their name indicates, non-market benefits are those which are not recognized in money terms, but are very valuable ethically, biologically and economically.

When we speak of the forest sector, we should consider that there are several groups or sub-sectors that use the same resource for different objectives. For example, there is the timber sub-group, the furniture sub-group and many other sub-groups. Nevertheless, in spite of the fact that different groups use the same resource, in general they do not have the same interests, responsibilities and commitments regarding this resource. There must be similar productive activities, shared interests and some formal or informal collective organization, as is the case of the sub-sector of timber producers, or the sub-sector of protected areas. In practice, the forest sector should be made up of all sub-sectors that are related to forest and timber resources. In a broad sense, the forest sector should be made up of all the economic agents which operate around the different environmental services and products of the forest, as described in Table 9.1.

Forest Resources in Costa Rica

As mentioned above, the forest has been recognized as a producer of lumber, firewood and cellulose. Since the prices of the first two have been relatively low and cellulose requires a much more complex industrial process supported by a large investment, society in general and economic policy in particular in Costa Rica have failed to recognize the social importance of the forest. Typically, the forest has been treated as an obstacle for activities which

Table 9.1 *Potential for forest resources*

Forest goods and services	Market benefits	Non-market benefits
1. *Timber*		
From primary production forests (approx. 180 000 ha)	Timber	
	Firewood	
Forest plantations (approx. 140 000 ha)	Utility poles	
Secondary forests (approx. 400 000 ha)		
2. *Non-timber*		
From primary production forests (approx. 180 000 ha)	Medicinal herbs	Genes
	Dyes	Chemical substances
Secondary forest (400 000 ha)	Ornamental plants	Odours/fragrances
	Seeds	
	Vines	
	Palms and construction material	
	Material for handicrafts	
3. *Maintenance of the water cycle*		
Water supply	Bottled purified water	Mineral springs
Urban	Water in tank trucks	Drinking water
Urban and rural		Water for other purposes: agriculture, industry, energy, transportation and tourism
Protection of agricultural lands		
Hydroelectric energy		
Loss of hydroelectric productivity		Water quality
		Watershed renewal and the water level in rivers

4. *Regulation of the micro-climate*		Local and horizontal precipitation
5. *Flood control*		Prevention of natural disasters
6. *Control of erosion*	Agricultural productivity	Soil conservation and fertility
7. *Control of sediments*	Production of energy	Water quality
8. *Water transportation*	Transportation of people, material and products	Transportation of the species
9. *Wind and noise control*	Prevent agricultural losses	People's enjoyment
	Permanence of the forest	Maintain the biodiversity
10. *Scenery*	Photography	Recreation
	Painting	
	Fishing	
	Videos	
	Tourism	
11. *Recreation and ecotourism*	Fishing and hunting	Recreation
	Visits to parks and tourism	
12. *Maintenance of resilience*		Health of the ecosystems
13. *Cultural and religious services*		Intergenerational transmission of culture and religion
14. *Preservation of the ecosystem and biodiversity*	Scientific tourism	Bioeducation
		Research
15. *Climate change*	Fixing carbon dioxide	Climate control

Source: Segura, O. *et al*. (1996), prepared by the author with contributions from CINPE-UNA and CCAB-AP colleagues.

generate a market-based income and economic development through mainly agricultural activities.

According to the General Forest Administration (*Dirección General Forestal*, DGF), this desire to increase income caused Costa Rica to change from having 72 per cent of its land in natural forests in 1950 to only 49 per cent in 1983, and 35 per cent in 1994. In the 1960s and 1970s this meant that approximately 50 000 hectares were deforested annually, with the consequent loss of many of the environmental services and products. This dramatic loss of forest has recently forced the country to orient its policies toward conservation and management of natural resources. In the late 1990s, Costa Rica was recognized internationally as a country that protects its forests as well as its flora and fauna. At the present time, Costa Rican forests may be classified in three categories: primary, secondary and plantation forests.

Primary forests

Primary forests (Segura *et al.* (1996)), defined as those that include virgin forests, are sub-divided into forests for preservation and forests for production. Approximately 25 per cent of the country's land is devoted to the conservation of ecosystems, under different legal regimes for their protection. The protected areas cover over one million hectares, distributed in nine conservation areas, and together these areas form the National System of Conservation Areas (*Sistema Nacional de Areas de Conservación*, SINAC). Since around 18 per cent of the protected areas are in private hands, efforts are being made to buy them and to provide for a sustainable management of the forest in the protected areas, by relating this to the activities of the neighbouring communities.

According to studies carried out by SINAC and by the Technological Institute of Costa Rica (*Instituto Tecnológico de Costa Rica*, ITCR), the area of the primary forests outside the protected areas is approximately 180 000 hectares. Around 60 per cent of these forests are concentrated in three areas which are used commercially. They are located in the regions of the Osa Peninsula, the Northern Zone and Baja Talamanca. The remaining 40 per cent are distributed in small areas and most of these forests are located within territories where indigenous communities live. The annual deforestation rate for 1994 was estimated by DGF to be 8000 ha, and for 1995 and 1996 it was 5000 per year.

Secondary forests

Secondary forest refers to forest areas that are regenerating naturally, after the disappearance of a previous forest. Generally, they are different in their composition from pre-existing forest species and they go through two stages prior to actually becoming forests. The first stage is commonly known as scrubland and the second as thickets.[4] According to the Promotion Department in SINAC, the National System of Conservation Areas (1995), it is estimated that in Costa Rica there are around 400 000 ha of secondary forests, as a result of lands which were abandoned. They were previously used for cattle ranching, but there is no official record at the present time. According to interviews conducted by Segura *et al.* and the study carried out by Rosalba Ortiz (1996), the most outstanding conditions which have favoured natural regeneration include the fall of beef prices on the international market and the elimination of subsidies and credit for cattle ranching activities. Nevertheless, Pedroni (1992) has stated that secondary forest is also developing as part of the agricultural cycle, since traditional farmers are accustomed to allowing the soil to regenerate for some years, leaving the land fallow and then using it once again for productive activities.

Plantation forests

Last, forest plantations are an important part of the national forest resources. According to SINAC, by 1995, 139 166.55ha had been reforested. In addition to reforestation due to incentives from the government, there are others which are financed exclusively by private enterprise (16 072.10ha), of which two of the most outstanding recent projects are Precious Woods in Guanacaste and Stone Forestal in the south. This reforestation together with the system of forest management with certificates (22 120.4ha), forest protection under Article 63 (31 147.20ha) and the areas which have been covered by the Certificate for Forest Protection System (22 199.8ha), give us an overall total of 214 634ha, which is broken down in the Appendix.

According to Segura (1997), it would appear that, in contrast to other Central American countries, the forest sector in Costa Rica depends more on plantations for timber production than on primary forests. As mentioned above, this is due to the fact that Costa Rica's remaining forest cover is already quite reduced and the plantations have continued to increase.

Forest resources in the national economy

Forest resources – including market and non-market forest goods and services – contribute greatly to the national economy. Unfortunately, the System of National Accounts does not reflect the interrelationship between the environment and the economy. Specifically the contribution of forest resources to the gross domestic product (GDP) first appears on the line 'Agriculture, silviculture, hunting and fishing'. In addition, the Sectoral Administration of Energy (*Dirección Sectorial de Energía*) keeps a record of alternative energy sources, within which firewood is included and continues to be an important source of energy. Secondly, the aggregate value corresponding to the industrialization of timber appears in 'Manufacturing industries and the exploitation of mines and quarries'. In the following two sections, it will be shown in detail how these contributions are recorded, but first we will examine some attempts to measure the remainder of the benefits which forest resources provide and which do not appear in Costa Rica's normal, permanent statistics.

For an idea of the forest's economic contribution to Costa Rica in terms of environmental services, we must make an ecological, economic valuation of those services, and at the present time very few valuation efforts have been made. One of the best known documents is 'Accounts Overdue: Natural Resource Depreciation in Costa Rica' (Solórzano *et al.*, 1991). Here it is estimated that the annual depreciation due to the destruction of the forests fluctuates between US$42 million and US$412 million, according to the 1984 exchange rate. To understand what those figures mean, we can imagine a loss, in 1984, of US$65 per person, or a depreciation of forest resources between 1988 and 1989 which was 36 per cent more than the service on the public foreign debt of Costa Rica.

Kishor and Constantino (1993) of the World Bank estimated that approximate values for the payment for environmental services of the primary forest in Costa Rica ranged between US$162.2 and US$333.7 per hectare per year. Another study specifically addressing the economic ecological value of the water in Costa Rica (CCT and CINTERPEDS, 1996) suggests that the protection and maintenance of the water cycle depend directly on the management and sustainability of forest resources. Thus far, although this seems to be obvious, the maintenance costs of this coverage have not been internalized in the price of water in any country. The cost of water continues to be equal to zero and the payments cover only connection services, maintenance of pipes, water treatment and drainage. The estimated value in this study, which includes the maintenance of the forest cover, fluctuates between 121.96 and 157.71 Costa Rican colones (at the 1995 exchange rate) per cubic metre of water.

More recently, in a study done by the Tropical Scientific Center (*Centro Científico Tropical*, CCT), *Valoración de los Servicios Ambientales de los Bosques de Costa Rica* (Carranza *et al.*, 1996), value ranges are established for the conservation of the primary forest in compensation for its services in carbon fixing and the protection of biodiversity. A payment of US$58 per hectare per year is recommended for the forest conservation certificate (*Certificado de Conservación de Bosque*, CCB) for these services. This study proposes that the payment would be reduced to US$48 for forest plantations and US$41.76 for secondary forests.

According to the literature reviewed, it is clear that to date there is little agreement on what methodology to use to establish the value for the environmental services of the forest, how much the payment should be, or how it should be made. The topic of the internalization of environmental costs and benefits is being debated nationally and internationally, but the process of coming to an agreement will be quite long. Some of these services, such as carbon fixing in the trees, are becoming 'products' or tradable services sooner than others. Probably it is here that we should apply some effort so that we can be creative and innovative, and thus benefit from comparative advantage by being the first in global negotiations on these resources. It appears that national politicians had this in mind when they passed Forest Law no. 7575 (16 April 1996), which defines environmental services and establishes the incentives for some of them. This change of perspective in economy, which has implied the creation of an economic policy favouring the environment and forest resources in particular, has been strengthened especially with the publication of some of the studies mentioned above.

SILVICULTURE

Timber and Improvements

As mentioned in the previous section, forest management and maintenance activities are not recorded separately from the rest of the agricultural activities. As a whole, cattle ranching, coffee, bananas, cacao, rice, corn, beans, cotton, tobacco, potatoes, cassava, plantains, onions, silviculture, hunting, fishing, poultry for meat, poultry for eggs, and others (included on the line 'Agriculture, hunting and fishing') contribute 19 per cent of the GDP, thus becoming the second most important activity in the national economy (see Figure 9.1).

The silvicultural activity composed of 'timber and improvements' represented, between 1990 and 1996, an average of 5 per cent of the Agricultural Gross Domestic Product (AGDP). According to the Agricultural Department of the Central Bank of Costa Rica *(Banco Central de Costa Rica*, BCCR), felling timber in the forest is recorded in the 'timber' section, and the section 'improvements' corresponds to the renovation or first-time planting (including living fences). The average participation of timber and improvements was 0.0085 per cent of the total GDP between 1990 and 1996. Its average participation in the AGDP in the same period was 3 per cent in timber and 2 per cent in improvements, as observed in Figure 9.2.

It is important to point out that silviculture increased its contribution in the AGDP from 1990 to 1996 by 92.6 per cent, going from 6 216 900 000 Costa Rican colones in 1990 to 14 211 600 000 in 1996 (see Table 9.1). However, this variation can be misleading if the data are not deflated at a constant value; in this sense, as seen in Table 9.2, since the figures were deflated at

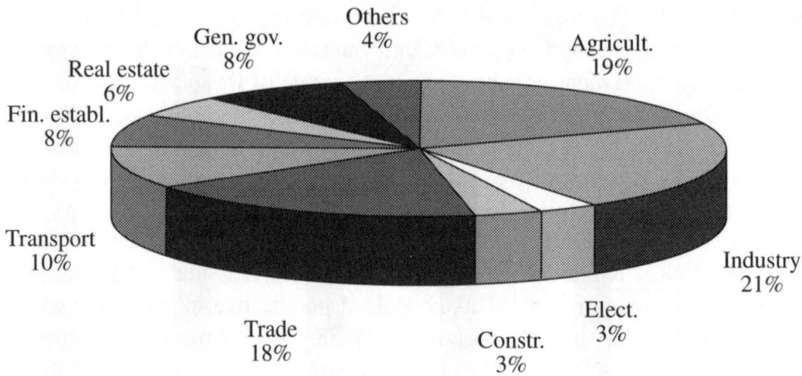

Source: Central Bank of Costa Rica (BCCR).

Figure 9.1 Participation of the different economic activities in the GDP in 1995 (in percentages)

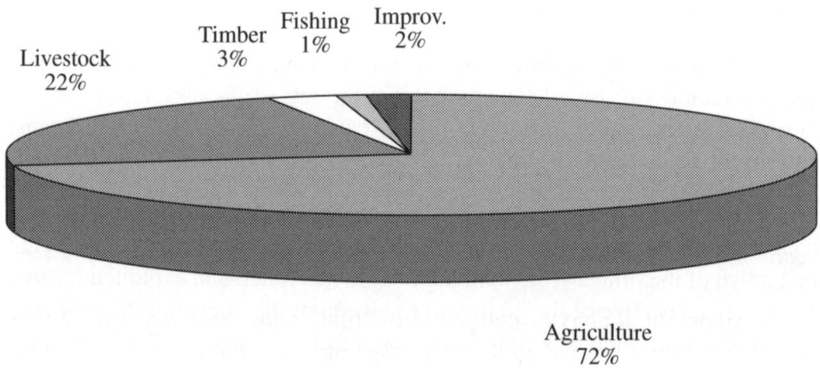

Note: The average is based on 1990–96.

Source: Central Bank of Costa Rica (BCCR).

Figure 9.2 Percentage of participation in the AGDP (average of 7 years)

1996 consumers' prices, this resulted in an important reduction in the participation. For 1990, 'timber' represented 78.45 million colones and in 1996 only 52.6 million; 'improvements' went from 46 million colones in 1990 to 43.7 million in 1996. These data may be observed both in Table 9.2 and in Figure 9.3.

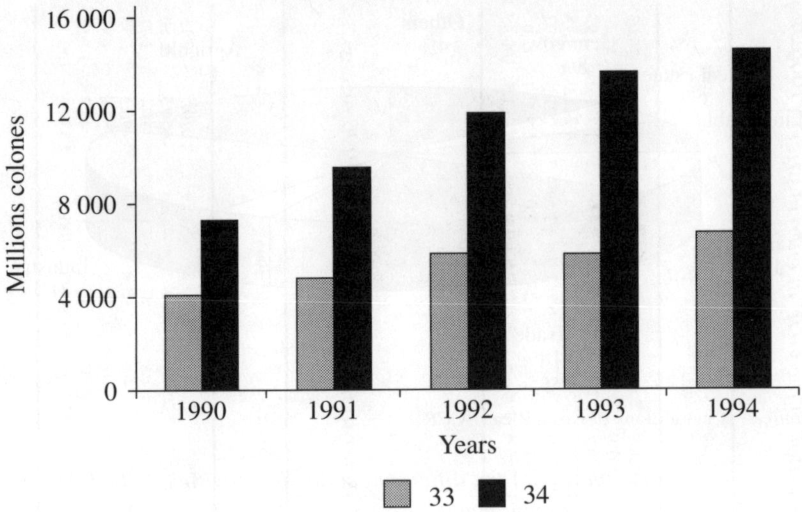

Source: Central Bank of Costa Rica (BCCR).

Figure 9.3 Aggregate value of branches 33 and 34 (millions of colones), 1990–94

Fuelwood

According to Table 9.4 it can be seen that, in the 1980s, fuelwood still contributed more than 30 per cent of the total energy production. This meant that much of the timber from national forests was felled and exploited for that purpose, irrespective of the quality of the wood or the impact of the process.

This situation could be said to have changed drastically from 1990, when the contribution of fuelwood to total amount of energy produced in Costa Rica was 30.38 per cent. By 1994 it had been reduced to only 12.69 per cent (Ballestero, 1996). In the same table other sources of energy, especially hydroelectric energy, and others including thermal energy, may be seen to have increased their production. It is also important to emphasize the case of vegetable waste, such as rice hulls and coffee skins, which can also be used as fuelwood, although these contributed only 4.87 per cent to the total 1994 energy supply.

THE TIMBER INDUSTRY

If we include the forest sector in industry and wood products such as furniture and paper (since records do exist for these), we can determine its share

Table 9.2 *Participation of silviculture in the contribution of the agricultural sector to the GDP at current prices (1990–96) in millions of colones*

Areas	1990	%	1991	%	1992	%	1993	%	1994	%	1995	%	1996	%
Agriculture	56 248.0	66.86	85 636.93	70.67	106 394.0	70.89	12 535	71.7	166 028.0	75.55	199 139.0	76.56	226 017.0	76.03
Livestock	19 094.0	22.70	23 837.32	19.6	30 468.8	20.30	34 314.0	19.6	35 801.0	16.2	42 200.3	16.2	4 528.0	15.2
Fishing	2 562.5	3.05	2 839.8	2.34	3 354.9	2.24	3 614.0	2.07	4 976.9	2.26	6 777.8	2.6	11 771.0	3.96
Silviculture	6 216.0	7.39	8 862.0	7.3	9 860.5	6.57	11 495.0	6.58	12 959.0	5.90	11 974.0	4.6	14 211.0	4.78
Timber	3 736.4	4.44	5 457.3	4.5	6 271.0	4.1	7 926.0	4.53	9 378.0	4.27	6 741.0	2.59	8 221.0	2.77
Improvement	2 480.5	2.95	3 404.7	2.8	3 589.0	2.39	3 569.6	2.04	3 581.0	1.6	5 233.2	2.0	5 990.0	2.0
AGDP	84 121.9	100	121 176.1	100	150 079	100	174 778	100	219 766.4	100	260 092.9	100	297 281.5	100

Source: Central Bank of Costa Rica.

Table 9.3 *Participation of silviculture in the contribution of the agricultural sector to the GDP at constant 1996 prices (1990–96) in millions of colones*

Areas	1990	%	1991	%	1992	%	1993	%	1994	%	1995	%	1996	%
Agriculture	1 693.64	71.61	1 789.68	71.17	1 887.5	72.18	1 924.7	71.90	1 973.7	71.60	2 074.8	72.29	2 070.5	72.13
Livestock	511.11	21.61	549.88	21.87	572.2	21.88	604.8	22.59	630.9	22.89	668.5	23.29	654.5	22.80
Fishing	36	1.52	32	1.27	30.9	1.18	28.3	1.06	30.4	1.10	31.1	1.08	49.1	1.71
Silviculture	124.45	5.26	143	5.69	124.3	4.75	119.2	4.45	121.4	4.40	95.8	3.34	96.3	3.35
Timber	78.45	3.32	93	3.70	79.9	3.06	79.2	2.96	85.4	3.10	52.6	1.83	52.6	1.83
Improvement	46	1.94	50	1.99	44.4	1.70	40	1.49	36	1.31	43.2	1.51	43.7	1.52
AGDP	2 365.2	100	2 514.56	100	2 614.9	100	2 677	100	2 756.4	100	2 870.2	100	2 870.4	100

Source: Central Bank of Costa Rica.

Table 9.4 *Costa Rica: annual domestic energy supply according to source and percentage of participation 1980–95 (terajoules)*

Year	Wood	Percentage of particip.	Vegetable waste	Percentage of particip.	Electric.	Percentage of particip.	Others	Percentage of particip.	Total
1980	18 079	29.77	5 288	8.71	7 072	11.65	30 287	49.87	60 726
1981	18 078	30.28	5 372	9.00	7 741	12.96	28 520	47.76	59 711
1982	18 136	33.32	5 183	9.52	7 457	13.70	23 651	43.45	54 427
1983	18 112	32.42	5 501	9.85	7 854	14.06	24 401	43.68	55 868
1984	22 660	35.19	6 738	10.47	8 661	13.45	26 326	40.89	64 385
1985	24 670	36.10	6 165	9.02	9 212	13.48	28 284	41.39	68 331
1986	23 854	34.05	5 989	8.55	9 945	14.20	30 263	43.20	70 051
1987	24 488	34.03	5 928	8.24	10 655	14.81	30 889	42.93	71 960
1988	23 992	32.16	6 198	8.31	10 906	14.62	33 502	44.91	74 598
1989	22 919	30.29	5 646	7.46	11 430	15.10	35 682	47.15	75 677
1990	23 765	30.38	6 207	7.94	11 977	15.31	36 273	46.37	78 222
1991	23 424	29.30	6 597	8.25	12 353	15.45	37 577	47.00	79 951
1992	23 726	28.19	4 086	4.85	13 191	15.67	43 166	51.28	84 169
1993	10 158	13.35	4 203	5.52	14 045	18.46	47 682	62.67	76 088
1994	10 440	12.69	4 010	4.87	15 204	18.48	52 608	63.95	82 262

Source: Prepared by author, based on data from *Memoria Estadística del Sector Energia de Costa Rica*.

of the exports, imports, the number of firms and the labour force required for that production. In this sense, first, we will present a description and a comparison, based on the Standard International Industrial Classification (SIIC), where branch 33 corresponds to industry and timber products including furniture, and 34 to the manufacture of paper and products, printers and publishers. Second, we will review in detail the different types of industrial firms included in branch 33, since in Costa Rica this branch is actually one which transforms timber products and there is no important transformation of timber products to be classified in branch 34.

Comparison of Branches 33 (Timber Industry and Products) and 34 (Manufacture of Paper and Products) within the Industry

As observed in Figure 9.3, the aggregate value of branches 33 and 34 increased for the period 1990–94, with a growth rate in branch 33 of 15.02 per cent, and 6.61 per cent for 34; this is a reflection of the nominal growth shown by the sub-divisions.

Nevertheless, we must point out that there is a significant difference in regard to the contribution of one branch or the other. Branch 34, the manufacture of paper and products, printers and publishers, is more important during the entire period studied than branch 33. In regard to the participation rate of these branches within the total aggregate value of the industry, there is a participation of 9.77 per cent for the period, going from 11.03 per cent in 1990 to only 8.71 in 1995. If both are broken down, branch 33 is the one with the lowest percentage; in 1990, this branch represented 3.92 per cent, and branch 34 was 7.12 per cent. In 1995, the former went down to 2.75 per cent and the latter to 5.96 per cent.

With respect to exports and imports in the same branches, as seen by comparing Figures 9.4 and 9.5, as well as in Table 9.5, it must be stressed that branch 33 showed a surplus for the entire period, whereas branch 34 had a deficit from 1990 to 1993. The statistics on the exports and imports of both branches for 1993 also reflect the deficit and surplus existing for each one of these branches.

In regard to the number of firms in these branches, it must be recognized that they represent relatively low percentages, with respect to the industry total. Branch 33 had 14.13 per cent and branch 34 had 8.24 per cent between 1991 and 1995. This comparison can be seen in Figure 9.6. Finally, Table 9.6, regarding the firms' location, reveals how the firms have tended to be concentrated near the capital of the country: the firms in branches 33 and 34 were mostly located in San José.

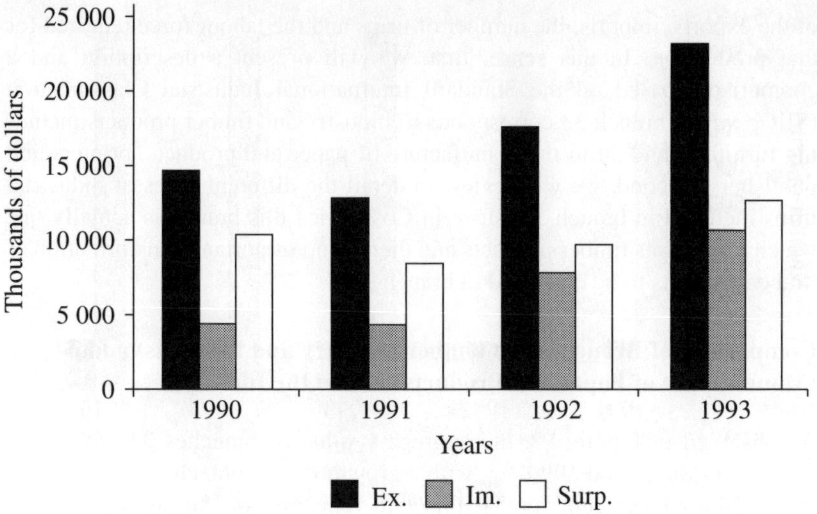

Source: General Statistics and Census Administration (*Dir. Gen de Estadísticas y Censos*).

Figure 9.4 Exports, imports and surplus of branch 33

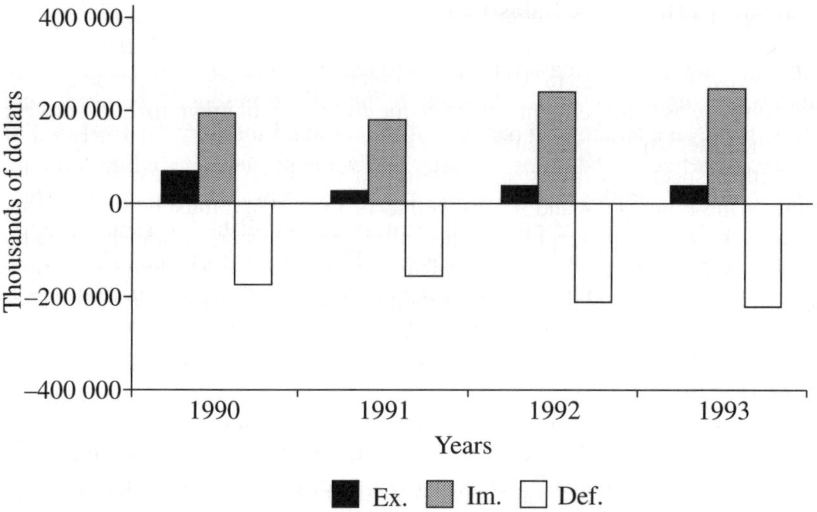

Source: General Statistics and Census Administration (DGEC).

Figure 9.5 Exports, imports and deficit of branch 34

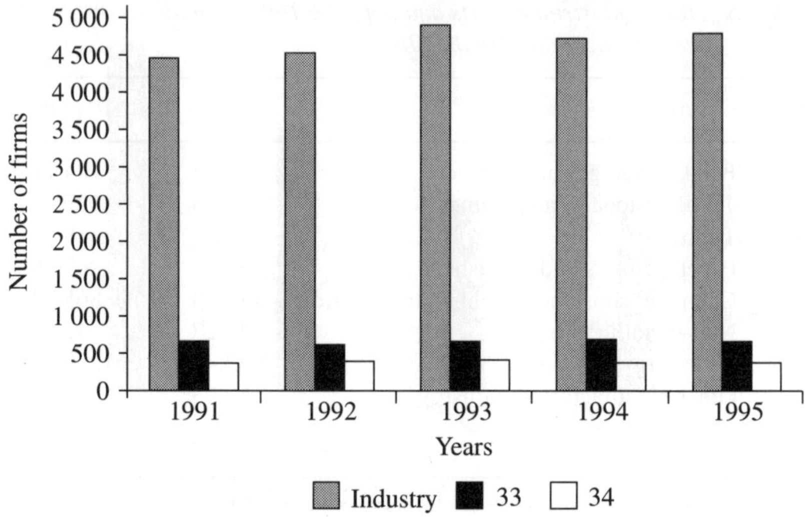

Source: Costa Rican Social Security Administration (CCSS).

Figure 9.6 Number of firms in the industry and in branches 33 and 34, 1994–5

Labour Market of the Industry

Other important information required to describe branches 33 and 34 is the number of employees in each. Thus branch 33 represents 5 per cent, and branch 34, an average of 6 per cent of the national industry for 1991–5. For 1995, in particular, the timber industry and wood products including furniture reported 6357 employees on the payroll, whereas branch 34 – the manufacture of paper and paper products, printers and publishers – reported 8305 employees. Detailed information for both branches is provided in Table 9.7. The comparison of employment in both branches is displayed in Figure 9.7. It is interesting to observe that, in general, the number of employees is higher in branch 34; however, as mentioned above, the number of firms is greater in branch 33.

With respect to the destination of industrial production, as observed in Table 9.8, in 1993 both branches sold a high percentage of their production to the local market; branch 33 dedicated 79.8 per cent, and branch 34 devoted 89.2 per cent of its production.

Table 9.5 Total industrial exports and imports: 1993 (in millions of dollars, according to the SIIC)

SIIC	Description	Exports	Imports
31	Food, beverages and tobacco	326.07	104.71
32	Textiles, apparel and leather	117.96	127.2
33	Timber	21.2	8.21
34	Paper, printers and publishers	33.2	263.76
35	Chemical substances, rubber and plastic	175.33	865.92
36	Non-metallic minerals	22.27	42.53
37–8	Metallurgy	132.4	1289.83
39	Other manufacturing industries	12.93	27.82
Total		841.36	2729.98

Note: The information of industrial exports does not include free zones and garment assembly.

Source: DGEC, Chamber of Industry, 1995.

Table 9.6 Costa Rica: location of firms in branches 33 and 34, by province: 1990–5

Provinces	1990	1995	Growth
San José	265	271	2.26
Alajuela	169	200	18.34
Cartago	74	69	–6.76
Heredia	63	69	9.52
Limón	22	30	36.36
Guanacaste	15	26	73.33
Puntarenas	17	25	47.06
Total	625	690	

Source: Costa Rican Social Security Administration (*Caja Costarricense del Seguro Social*, CCSS).

Table 9.7 *Number of employees in the industrial branches involved in timber (in December of each year)*

SIIC		1991	1992	1993	1994	1995
	Industry total	115 045	123 344	140 132	138 821	135 884
33	Timber industry and products including furniture	6 248	8 468	7 378	6 789	6 357
331	Timber and cork industry and products except furniture	3 906	6 097	4 140	4 099	3 711
3311	Sawmills, finishing shops and other shops	3 412	3 451	3 851	3 819	3 356
3312	Manufacture of wood and cane containers and small cane articles	0	0	4	7	20
3319	Manufacture of wood and cork products (avg)	494	275	285	273	335
332		2 342	2 371	3 238	2 690	2 646
3320	Manufacture of furniture and accessories (not metal)	2 342	2 371	3 238	2 690	2 646
34	Manufacture of paper and products, printers and publishers	7 313	7 657	8 549	8 123	8 305
341	Manufacture of paper and paper products	2 966	3 023	3 092	3 175	3 398
3411	Manufacture of pulpwood, paper and cardboard	70	80	79	214	360
3412	Manufacture of paper and cardboard containers and boxes	2 840	2 902	2 951	2 258	2 341
3419	Manufacture of pulpwood, paper and cardboard (avg)	56	41	62	703	697
342		4 347	4 634	5 457	4 948	4 907
3420	Printers, publishers and related industries	4 347	4 634	5 457	4 948	4 907

Source: CCSS.

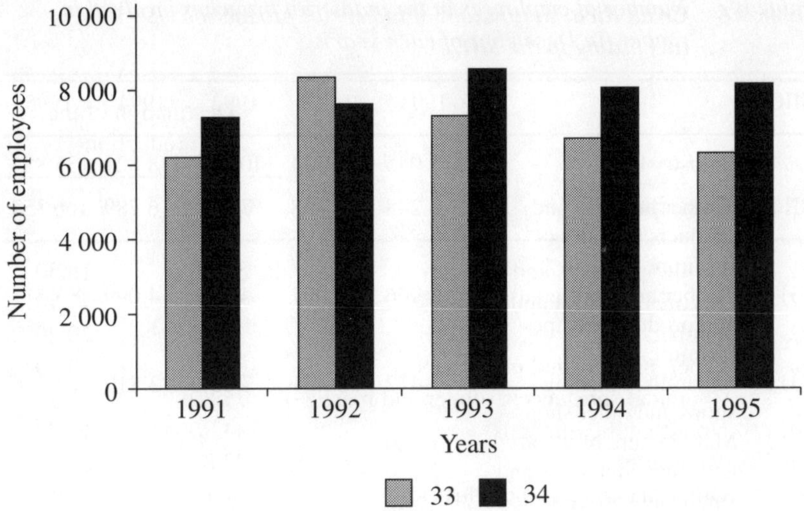

Note: Other institutional sources, aside from the Costa Rican Social Security Administration (*Caja Costarricense del Seguro Social*, CCSS) and literature on the sector provide somewhat confusing information, and mention the lack of studies on informal employment in the sector. Some of the informal employment can include basically unskilled workers, but in other cases there are professionals such as forestry engineers and consultants who are not recorded on the payroll as salaried workers, or in any other statistics. The forest study (*Diagnóstico Forestal*, UICN/IUCN, 1996) states that the last forest census was published by the DGF in 1988, and reported that in 1986 there were 2214 employees in the sawmill industry. This study also includes data such as the total amount of direct employment in the sector (21 000) as well as the number of forestry engineers in the public sector (199), in research and training centres (47) and in the private sector (100).

Source: Costa Rican Social Security Administration (CCSS).

Figure 9.7 Number of employees in branches 33 and 34, 1991–5

Description of Industrial Branch 33: Timber

Timber industry

The timber industry is defined as the work involved with the owner of the forest – who sells the standing timber – and the forest industry which buys it (UICN/IUCN, 1996). In other words, this type of industry includes felling the standing timber, dragging it to the place where it will be loaded for transportation, loading the timber truck, transporting it to the sawmill and unloading the timber there. As described in a number of studies about the forest industry, (ibid.) in this stage timber prices are relatively low, legal dispositions are unstable and inefficient and the future is uncertain. These same conditions

Table 9.8 Costa Rica: destination of industrial production, 1993 (according to the SIIC)

		Destination of the production	
SIIC	Description	% Local	% foreign
31	Food, beverages and tobacco	81.70	18.30
32	Textiles, apparel and leather	52.30	47.70
33	Timber	79.80	20.20
34	Paper, printers and publishers	89.20	10.80
35	Chemical substances, rubber and plastic	75.80	24.20
36	Non-metallic minerals	84.30	15.70
37–8	Metallurgy	72.8	27.20
39	Other manufacturing industries	n.d.	n.d.
Total		77.90	22.10

Source: BCCR, DGEC, Chamber of Industries, 1995.

cause timber dealers who are in charge of this process to have a short-term perspective and to prefer ready cash-flow.

In addition, the owners of standing timber often are not able to industrialize timber any other way. Thus historically they have distributed 'free' or extremely cheap wood, in exchange for having their lands cleaned off or canals dug on their farms. This situation of monopsony is very difficult for small farm owners to avoid. In general, for farm owners it is very clear that alternative activities are much more profitable than preserving the forest. This is mainly due to the fact that the positive externalities of conservation and handling are not taken into consideration in favour of the forest sector, just as the negative externalities are not taken into account with regard to the other alternative activities which are not included either in the market prices of their products.

Stewart's study (1992) which is also mentioned in the *Diagnóstico Forestal de Costa Rica* (UICN/IUCN, 1996), adds other problems such as the costs of getting a permit to use the forest legally, the prohibition of the export of logs which affects the prices of timber negatively, and other costs such as taxes, deposits for guarantees, and so on. In the end, they make the timber business (the timber industry) profitable only if timber is exploited in general, instead of making a forest management and utilization plan which would imply the protection of certain parent trees and shade trees.

This situation may have changed somewhat due to the passage of a new forest law (Article 28) which eliminates the necessity of permits for the felling, transportation, industrialization and exportation of timber from plantation forests. Likewise, according to Segura *et al.* (1996) important progress has already been made in the situation of the domestic timber market and, concretely, this is the incorporation of standing timber in the agricultural products market (*Bolsa de Productos Agropecuarios*, BOLPRO). Both in this publication as well as in the declarations of the Minister of the Environment and Energy (Ministry of Environment and Energy, *Ministerio del Ambiente y Energía*, MINAE), it has been suggested that with this system the intermediary is eliminated, and since the owner negotiates directly, he obtains almost twice as much. This increases the profitability of the forest for the owner, enabling him to compete with other uses of the land. At the same time, it is also beneficial for those who process the timber. The purchase of standing timber allows it to be felled according to demand. In addition, BOLPRO will enforce the management plans approved by MINAE and conservationist organizations can buy trees so that they will not be felled, thus favouring the possibility of creating a market for environmental services. In Costa Rica the first negotiation of management plans in BOLPRO was carried out in November 1995 and the second in February 1996.

Sawmill industry

In 1980 there were 204 sawmills; in 1986, 161. Data from the General Forest Administration *(Dirección General Forestal*, DGF) indicate that at the present time there are 185 sawmills or firms devoted to milling, which generate the largest amount of employment in the sector. However, according to several authors, such as Segura *et al.* (1996), they are characterized by their inefficiency, their generally limited capacity, their use of old machinery and obsolete technology with low performance. Other DGF data show that from 30 per cent to 40 per cent of the log is used; the rest is wasted or used for secondary products such as sawdust or wood shavings.

Under these conditions it is obviously impossible for the sawmill industry to compete with foreign firms. A MIEC–CCF (1995) study states that

> in this sector the combination of the different policies promoted leads to a residual productive process, which is not very competitive, efficient or inclined to modernization. The relative prices are highly distorted, thus significantly limiting the total use of the existing timber potential and making the owner of the forest decide whether to choose activities with a higher private profitability – as a product of the distortion – or to keep the forest as an activity itself.

Last, mention must be made of the lack of a different perspective for development, together with the need for funds to make changes. Schools and training programmes in higher education would be required to support and stimulate silvicultural activities. All of the above cause the forest sector and this industry in particular to be facing serious difficulties for development.

One industry which is very much related to sawmills is the construction industry. Again this industry is characterized by a low degree of development and a lack of standardization in the requirements. This has important consequences in the forest industry, since construction consumes around 60 per cent of the total timber production. In the first half of the 1990s, construction grew very rapidly in Costa Rica, and so did the credit assigned to it. Nevertheless, the sawmills did not improve their technology or management. The *Diagnóstico Forestal* (1996) mentioned above, suggested that there is a high percentage of untrained personnel in sawmills and that the ratio of administrative personnel to workers is 1 to 1.83, whereas it should be 1 to 8.

Industry of gameboards, match and toothpick factories

According to the *Diagnóstico Forestal* (1996), there are only two plywood factories, one of pressboard and one of veneer. It also states that the pressboard factory is the most underused, since it takes advantage of waste from plywood factories and there is not enough waste available. However, according to interviews which we have had recently, one plywood factory works 24 hours a day, with four shifts, and is buying waste wood from other national sawmills as well as using material from pruning at plantation forests.

On the other hand, two match factories and a toothpick factory need standard high-quality white wood for their products. The former operate at an acceptable level of their installed capacity (85 per cent), but the latter is only operating at 54 per cent of its capacity.

Secondary timber industry

The secondary industry is made up of 404 production units, including furniture factories, small and medium furniture-making shops and a few craftshops. They are mostly family-oriented microenterprises which are located in the Central Valley. There is a limited number of employees in each factory and the design and construction in solid wood predominates; there is a lack of knowledge of advanced techniques utilized at the international level.

Very few industries are vertically integrated with the forest and its different links. The department for the integration of forest and industry which existed in the former DGF disappeared some time ago. In addition, in many cases this integration is conceptualized as a task which the owner of the company should perform, establishing a number of links related to production. Thus only medium and large firms could establish, maintain and exploit a forest chain. This type of integration has been possible in firms such as PORTICO, which manages its own forest and exports doors and windows to the United

States and Europe. Something similar occurs with a few other firms, such as El Buen Precio S.A. However, the integration of forest and industry should be conceived as a possibility for alliances between some reforestation planters or forest owners and sawmills and furniture makers, thus forming a strong chain of cooperation between the different businesses.

The few existing export industries of timber products benefit in several ways. For example, one benefit is timber prices which are lower than those of the international market. Another is the national incentives for export industries (such as the tax bond certificates – *Certificados de Abono Tributario*, CAT), and last, access to profitable international markets, especially when dealing with products which are certified for good forest management, as is the case for the doors exported by PORTICO S.A.

Trade of forest products

The balance of trade of forest products is positive, but there is an enormous difference between the sub-branches that make up the total (see Tables 9.9 and 9.10). The branches of doors, windows, veneer, pressboard, shoe lasts, ladders, handles, hangers and rods have a trade surplus, but the export of doors is one of the most dynamic products. The table on imports shows how plywood is one of the most frequently imported forest products in Costa Rica, although the country also exports a great deal of the same product.

Table 9.9 FOB value of export forest products, 1990–93 (dollars)

Products	1990	1991	1992	1993	Total
Handicrafts	4 260 805	1 077 601	1 418 061	5 805 597	12 562 064
Toilet seats	565 704	548 097	405 250	394 284	1 913 335
Sawdust	0	3 048	4 312	0	7 360
Bamboo	0	0	6 554	163 207	169 761
Wooden spools	194 396	9 021	259 677	0	463 094
Composition board	1 867 460	2 590 700	3 685 657	7 655 400	15 799 217
Plywood	1 707 908	1 963 898	2 791 712	3 574 740	10 380 258
Sheets of wood	1 078 864	1 042 321	1 109 892	0	3 231 077
Manufacturing	1 360 354	1 700 664	2 226 406	1 768 379	7 055 803
Furniture	5 073 914	4 054 213	3 758 883	1 628 675	14 515 685
Panelling	722	377 576	0	0	378 298
Doors	5 927 621	3 344 876	6 061 229	7 218 885	22 552 611
Milled timber	18 987	108 410	846 452	6 231	980 080
Total (m3):	22 056 735	16 820 425	22 574 085	28 215 398	89 666 643

Note: FOB value is value of the export product in dollars.

Source: SINAC, MINAE, 1996.

Table 9.10 CIF value of imported forest products 1990–93

Products	1990	1991	1992	1993	Total
Sawdust	271	7 676	41 014	0	48 961
Charcoal	1 384	13 641	0	0	15 025
Milled timber	122 238	28 379	11 993	59 298	221 908
Plywood	2 887 854	109 619	2 888 301	818 409	6 704 183
Wooden mouldings	25 596	78 107	8 223	0	111 926
Manufacturing	220 177	119 572	204 357	2 279 017	2 823 123
Furniture	51 157	35 382	8 486	423 634	518 659
Panelling	107	21 535	11 993	100	240 528
Broom handles	98 119	46 942	219 712	0	364 773
Utility poles	74 700	353 630	344 112	98 310	870 752
Doors	10 447	59 455	0	397 721	467 623
Total (m3):	3 598 943	873 938	3 738 191	4 176 389	12 387 461

Note: CIF value is value of the import product in dollars including the exit port in the original country.

Source: SINAC, MINAE, 1996.

Moreover, between 60 per cent and 80 per cent of the input for the production of paper and cardboard products is imported and in 1993 the furniture branch had a trade deficit of more than US$ 2.0 million. It should be emphasized that the import and export figures of cardboard do not reflect reality entirely, since the cardboard boxes manufactured in Costa Rica are used mainly for export products, but obviously they are not recorded as exports but as part of the cost of export products. The export of bananas (over 70 million boxes per year), for example, is one of the agricultural activities which uses a lot of cardboard boxes. With this we wish to stress that, when analysing irregularities in the balance of trade of forest products, their contribution to the generation of value in other national products must also be taken into account.

Between 1986 and 1993, the export of forest products grew at an annual accumulative rate of 20.83 per cent, mainly due to the incentives created in the mid-1980s for the export of non-traditional products. However, the inefficiency of the industry, together with the high prices of raw material and certain production factors, caused the prices of furniture on the national market to be higher than on the international level. Consequently, furniture makers reduced their exports from US$5 073 914 in 1990, to US$1 628 675 in 1993 (Table 9.9). An increasing number of shelves, tables, wardrobes, bookcases and other similar products (often of a modular style) are being imported and compete on the national market with lower prices and more attractive designs.

Economic globalization has begun to affect the forest industry and, according to the president of the Costa Rican Chamber of Forestry (*Cámara Costarricense Forestal*, CCF), this could eliminate from the activity many of the small industries since they are not prepared technologically or financially to compete at that level. Thus the forest industry and the government are making efforts to modernize the sector. Therefore, in May 1995, an agreement was signed between the Ministry of Industry, Economy and Trade (*Ministerio de Industria, Economía y Comercio*, MIEC) and the CCF, by means of which a Modernization Strategy for National Industry will be implemented to reconvert the timber industry. The main actions of the strategy have been summarized (Box 9.1) by Segura *et al.* (1996):

BOX 9.1 ACTIONS FOR AN INDUSTRIAL MODERNIZATION STRATEGY IN TIMBER

Prepare a current assessment of the situation of the forest sector.
Design a strategy for industrial modernization.
Establish a sectoral labour committee.
Initiate an industrial reconversion programme for the forest sector.
Expedite the necessary procedures for the forest industry to have access to economic and financial resources.
Promote mechanisms for technical aid and training in the timber industry.

REFORESTATION INCENTIVES[5]

In Costa Rica fiscal incentives for reforestation have existed since 1979. According to the official information of the National System of Conservation Areas (*Sistema Nacional de Areas de Conservación*, SINAC), and as observed in the table in the Appendix, in 1995 there were 139 166.59ha (adding all reforestation programmes) reforested, 22 120.40ha managed and adding CPB and Article 63, 53 347.01ha protected. The first incentive applied was an income tax deduction which mainly favoured natural and legal persons who had to pay taxes. Then in 1988 incentives were introduced to increase the participation of medium and small producers – who did not necessarily have to pay taxes, but who were interested in reforesting. Last, around the end of 1995, with the new forest law, other incentives were created to compensate the owner of the forest for part of the environmental services provided for the national community. This incentive is called the forest conservation certificate *(Certificado de Conservación del Bosque*, CCB). This new forest law also introduces other incentives such as the exoneration of property taxes, exemption from payment of the tax on assets and protection against invasion of properties which are voluntarily submitted to the forest regime and/or devoted to forest activities.

The main fiscal incentives which favour the forest sector are summarized below.

1. Income tax deduction: this was established as the country's first forest incentive. The government promotes plantation forests to obtain the raw material which the country needs to prevent the destruction of the natural forest. Although this incentive was intended to be extensive and motivate land owners, most of Costa Rica's rural population was excluded since only large land owners pay income taxes.
2. Forest bond certificate (*Certificado de Abono Forestal*, CAFc): the 'current' forest bond certificate was established in May 1986. Its objective was to distribute democratically the resources which the government allocates to forest activities. The CAFcs are tax-exempt nominal titles with which any type of tax may be paid.
3. Advance forest bond certificates (*Certificados de Abono Forestal por adelantado*, CAFa): this incentive was created in 1988 in response to the problem caused by the CAFcs, as an incentive for small landowners who were unable to invest individually.
4. Forest management bond certificate (*Certificado de Abono Forestal para Manejo*, CAFma): this is a forest incentive whose main objective was to achieve a comprehensive management of the forests, except plantation forests. The CAFma is a title of nominal value in national currency, which may be traded or used to pay any type of national or municipal taxes or tariffs or any other type of tribute.
5. Fund for municipalities and organizations: Forest Law #7032 establishes a tax on forest activity and allocates 20 per cent of that collected from the exploitation of timber in the area to regional organizations and municipalities with the objective of implementing reforestation projects, managing watersheds, establishing nurseries, promoting the extension and development of the forests and constructing infrastructure.
6. Forest protection certificate (*Certificado para la protección del Bosque*, CPB; or CAFMA 2000: this is an incentive which was developed in 1994, consisting of granting residents of rural areas 10 000 colones per ha/year for five years. Any land (between one and 300ha) qualifies for this type of incentive if it is covered with natural forest and if the property owner agrees not to exploit it in any way for production, except for ecotourism. Thus the owners of primary forests are recognized for

the different environmental services provided to the national and international community.

7. Forest conservation certificate (*Certificado para la Conservación del Bosque*, CCB): the Forest Law of February 1995 established the forest conservation certificate with the objective of compensating the owner for environmental services which are generated from conserving the forest as long as no timber has been exploited for two years prior to the application and during its effective period, which cannot be less than 20 years. At the present time SINAC is preparing a proposal for the implementation of this incentive which will, by law, be financed by one-third of the funds from the 1995 increase in fuel prices.
8. Soft credit: another financial mechanism created for the development of forest activity is soft credit, which was established in 1983. The government, by means of the National Banking System – jointly with the programme known as COREMA AID-032 – established credit with an 8 per cent interest rate, with a 10-year grace period. It defines periods of as long as 30 years, depending on the species used for reforestation. Trust 340 MIRENEM-BNCR (National Bank of Costa Rica – *Banco Nacional de Costa Rica*), consists of all the assets, liabilities and equity of the final results of Trust 178 (AID-515-T-032) which is allocated to the Peninsula of Nicoya and to the counties of Puriscal, Acosta and Turrubares; Trust 04-87, with the Cooperative Bank (*Banco Cooperativo*, BANCOOP) is intended to finance small reforesters in the southern zone. It is composed of 30 per cent of 80 per cent of the income from the general forest tax, and 30 per cent of the legacies and donations which the forest fund accepts. Trust 19-91 FDF, Multi-donor Addendum Trust No 19-91 MIRENEM-BANCOOP R.L. It gets legal support particularly from the agreement for the administration of the investment resources of the project established with FINNIDA-MIRENEM-BANCOOP R.L. It is oriented toward small and medium producers.

   Legally constituted, recognized organizations which may legally exercise their rights and contract obligations qualify for soft credit if at least 70 per cent of their associates are small or medium producers which are participating in the project. The organizations must prove that they possess an efficient technical, administrative and accounting system. In addition, they should be willing to accept and apply the technical and administrative recommendations which BANCOOP and SINAC make. They should also prepare the project which will be financed, as well as having technical forest assistance.

To close this section on forest incentives, it is worth mentioning that opinions vary as to the achievements made with this national effort. Some believe that most plantations were a failure since it is supposed that each reforested hectare should produce approximately 25 cubic metres per year, and that, having reforested nearly 140 000ha, we should have had more than enough for the national timber market, which requires approximately one million cubic metres. Nevertheless, other analysts suggest that, although the plantations were not entirely successful, it must be considered that some, especially many small and medium producers, did not plant to take timber to market, but with a more comprehensive, and perhaps long-term, perspective. They could be more interested in the rest of the environmental services which the plantations, and the forest in general, produce. They may have the trees reserved like money in the bank, which will be spent in the future when their needs – or others' – require it.

SOCIAL ACTORS AND INSTITUTIONS

The government is one of the main actors in the forest sector with the role of planner, along with the municipalities, local governments, NGOs, the rural population, private enterprise, women, universities and others. Each actor has different interests, and even within each of the groups there is no unified position. How are these conflicting interests to be resolved?

During the current administration, the government has significantly modified its role as 'watchman' and controller and has assumed a more active position in the management of forest resources. Although this change is just beginning, it is based on the transformation of the entities which are responsible for the forest resources; that is, DGF, the Wildlife Administration *(Dirección de Vida Silvestre)* and the National Park Service (*Servicio de Parques Nacionales)*. These three bodies were dissolved and in their place the National System of Conservation Areas (*Sistema Nacional de Areas de Conservación*, SINAC) was established with the responsibility of contributing to and orienting the policies and the development of natural resources in all nine conservation areas composing it. SINAC is part of the Ministry of Environment and Energy (*Ministerio de Ambiente y Energía*, MINAE) which stimulates the active participation of NGOs and society in general. The municipalities are another governmental body which participates on a local level and according to the new law they are largely responsible for local forest resources.

There are also quite a few non-governmental organizations (NGOs) related to forest resources; for example, the National Institute of Biodiversity (*Instituto Nacional de Biodiversidad*, INBio) which does biodiversity prospecting, the Cooperation of the Forest and Timber Sectors (*Cooperación de los Sectores Forestal y Maderero*, COSEFORMA) which functions mainly in the northern zone of the country, and the Foundation for the Development of the Central Volcanic Range (*Fundación para el Desarrollo de la Cordillera Volcánica Central*, FUNDECOR) which works both in conservation areas and in buffer zones in that region (also the first NGO negotiating the sale of the capture of carbon dioxide). The Neotropical Foundation (*Fundación Neotrópica)*, the National Park Foundation (*Fundación de Parques Nacionales)*, the Costa Rican Association for the Conservation of Nature (*Asociación Costarricense para la Conservación de la Naturaleza*, ASCONA) and other smaller foundations are working in forest conservation.

The centres of research and training include the Tropical Agronomy Center for Research and Instruction (*Centro Agronómico Tropical de Investigación y Enseñanza*, CATIE); the National University of Costa Rica (*Universidad Nacional*, UNA); the University of Costa Rica (*Universidad de Costa Rica*, UCR); the Costa Rican Institute of Technology (*Instituto Tecnológico de Costa Rica*, ITCR); and the State Distance University (*Universidad Estatal a Distancia*, UNED). All of these offer research and training related to the forest sector. The University for Peace (*Universidad para la Paz*, Upaz), the School of Agriculture of the Region of the Humid Tropics (*Escuela de Agricultura de la Región del Trópico Humedo*, EARTH) and the Central American Institute of Business Administration (*Instituto Centroamericano de Administración de Empresas*, INCAE), also offer excellent training in agriculture and forestry. In addition there is the Organization of Tropical Studies (*Organización de Estudios Tropicales*, OET) which belongs to an international university research consortium.

The private groups are organized by trades, as follows: the Costa Rican Chamber of Forestry (*Cámara Costarricense Forestal*, CCF) to which the most important forest firms in the country belong; the National Chamber of Timber Industrial Firms (*Cámara Nacional de Industriales de la Madera*, CANAIMA), which includes approximately 50 associates; and the Rural Population's National Forestry Board (*Junta Nacional Forestal Campesina*, JUNAFORCA), which covers the entire rural population related to the forest sector.

CONCLUSION

Main Constraints on the Development of the Forest Sector

One of the main restrictions for the sustainable development of the forest sector is that national policies were concentrated in the political, financial and institutional support of other sectors such as those of agriculture, cattle ranching, non-traditional products and, lately, tourism. This emphasis on other sectors has promoted inappropriate customs, habits and routines which do not stimulate sustainable forest use.

Technological development has received little financial and research support and the packages applied have not improved the profitability of forestry. This situation is made worse by the reduced administrative and management capacity of the different links in the forest sector. There is very little integration between forest and industry, and even in industry there are very few experiences of integration and alliances. The development of forest activities has not taken into account the intellectual, technical, cultural and economic contribution of indigenous communities and women, which could strengthen it and make it competitive.

The application of forest incentives, in spite of their importance in the development of the reforestation of the country, has not taken advantage of the opportunity to motivate the rural population to take a more active role in the forest chain. In the entire procedure there is a lack of technical assistance concerning the exploitation, storage and marketing of timber articles and non-timber products. Moreover, although timber production by means of reforestation with incentives was not a complete failure, it can be said that it was not a total success either, and further planning would have been required for it to become an axis of national development.

Finally, environmental services are positive externalities for any forest, including the Costa Rican forests. In some cases no one knows about the services, and in others the techniques used for their valuation are still in initial stages and are being debated internationally. In this sense, the predominant view of the use of natural resources is biased toward agricultural production. In addition, the forest sector is facing problems of identity and of competition from other institutions. Nevertheless, we cannot – and must not – ignore the fact that in Costa Rica efforts are being made to incorporate environmental services in the sphere of the economy. One is the sale of fixing carbon in the trees, and another is the incorporation of regulations regarding environmental services in the recently passed Forest Law.

Potential for the Development of the Sector

Research is needed in order to study the competitiveness of the forest sector which would clarify the different links in the forest chain, and its capacity and potential as well. In this sense, if we began to visualize each one of the links related to the goods and services of the forest as elements which can be grouped in each industrial area, we would have clusters which would permit the development and promotion of competition at each of the links which it is composed of; and that would eventually increase competitiveness.

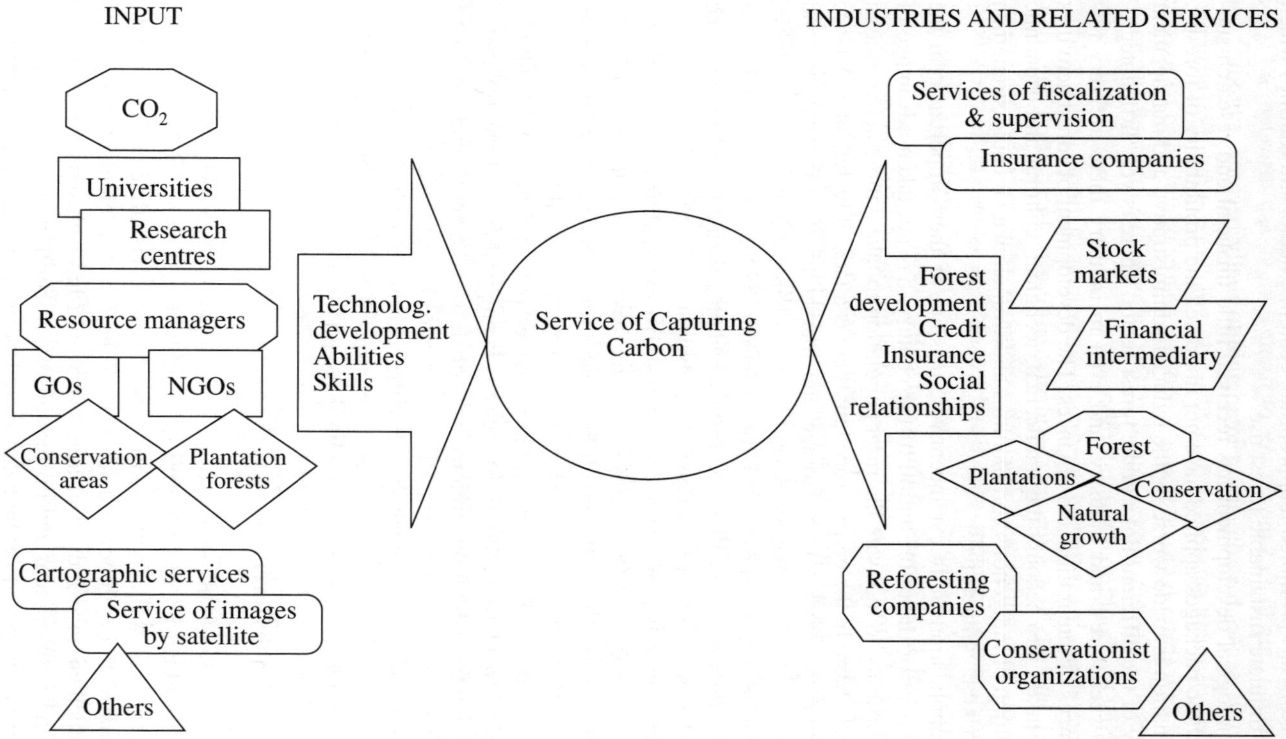

Figure 9.8 Economic group for the service of capturing carbon

We can take, for example, within this block or 'cluster' of activities of the forest sector only that of the service of fixing carbon, as illustrated in Figure 9.8. In this sense, for something intangible to really become a product (or commodity), many services or related work will be required. For example, we need CO_2 (carbon dioxide) as a generating source of carbon to be fixed in our forests. Different governmental (GOs) or non-governmental organizations (NGOs) should manage natural resources, in such a way that the GOs will direct their efforts toward the conservation areas (national parks, wildlife refuges), and the NGOs toward the plantation forests. In addition, the cooperation of universities and research centres is required to contribute to technological development, advice and technical assistance in this new field. All of these needs or inputs will provide the service of capturing carbon with technological development, abilities and skills.

Services of supervision of natural and plantation forests are part of the related industries and services. In addition, insurance companies are required as the entities which will be in charge of insuring the new 'product' stored in the trunks of trees in case of fire or loss. In addition, there are stock markets and financial intermediaries which are responsible for finances. Forestry engineers, biologists and other professionals should be in charge of the process of managing the forest, the plantations, natural regrowth and/or the conservation according to the particular case. Last, reforesting companies and conservationist organizations are part of the entities interested in this process. The above industries and related services provide the service of carbon capture with forest development, credit, insurance and social relationships.

Now all these requirements for the realization of a new forest product should be organized as well as possible so that there will be a real production of wealth and benefits for many, and so that Costa Rica will have a competitive place in the international market. How can this be achieved? What would be the best type of organization and production system, labour and interrelationship between the different links of the new chain of the forest cluster? These and many other questions must be answered by means of economic, ecological and institutional studies on the forest sector.

Another example of research could address the timber industry more specifically. To illustrate, we could concentrate on what we will call the Economic Group for the Timber Industry, as shown below. To examine this group or block we should take into account that, although we have sufficient input to produce timber, it is also indispensable to have services for felling, transporting the logs to secondary sites, transporting them to loading zones, loading and unloading, not to mention the capacity to drag them. The participation of workers, technicians and administrators is also required.

Sawmills, kilns, finishing, and the use of waste such as sawdust, kindling and wood shavings, are part of the related industries and services. The timber

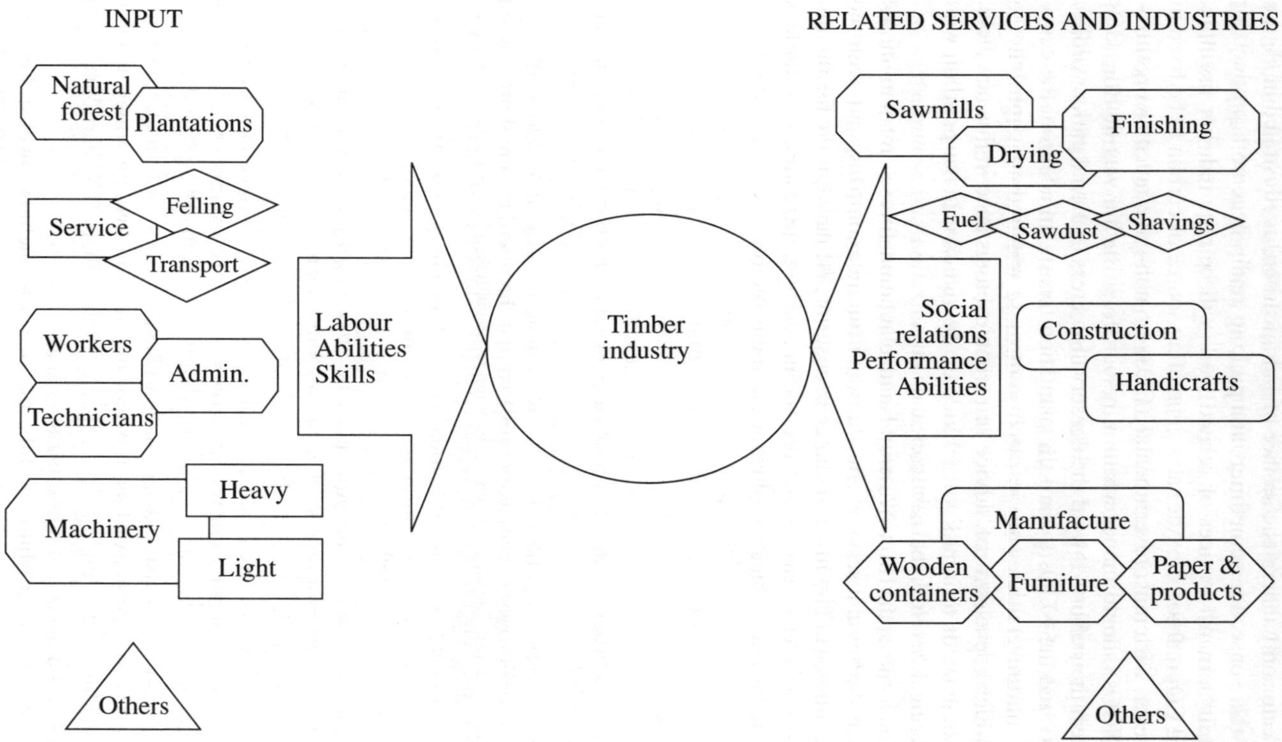

Figure 9.9 Economic group for the timber industry

industry can be useful in construction and secondary industries such as cabinet making and handcrafts. Another branch of the industry could include the manufacture of wooden containers and packing material, as well as paper and paper products. All of these different links in the timber industry are illustrated in Figure 9.9.

However, what must be emphasized here is that the sources of competitiveness of the industry as a whole emerge from the innovating forms of relationships established by the different sub-sectors and the different groups of individuals which carry out each task. These interrelationships are not new for the industry, but they are in the case of those which are emerging in the block which represents the service of carbon capture. However, what *is* new in the analysis of this traditional block of the industry is the possibility of discovering what may be new forms of organization and relationships between each one of the links. We must become as competitive as other countries that have the same resources and the same interrelationships, and even the same technology. The institutional relationships and the systems for the development of the activities and for adding value to the primary resources should be the main objective of research in this sector.

NOTES

1. The paper on which this chapter is based was produced under a framework of joint research cooperation between the International Center on Economic Policy for Sustainable Development (CINPE) of the Universidad Nacional in Costa Rica and the IKE group of Aalborg University in Denmark. The author would like to thank the Swiss Cooperation for Development (COSUDE) and Centro Agronómico Tropical de Investigación y Enseñanza (CATIE) for their support. The same document is going to be published in the SUDESCA working paper series at Aalborg University, Denmark.
2. For further information on the discussion of how a global perspective influences land use and policy-makers, see Gottfried (1995).
3. This paper has been made possible thanks to the cooperation between the International Center on Economic Policy for Sustainable Development (CINPE) of the Universidad Nacional in Costa Rica; the Swiss Cooperation for Development (COSUDE) through *Centro Agronómico Tropical de Investigación y Enseñanza* (CATIE) and the IKE group of Aalborg University in Denmark. I also wish to thank Jessica Castro and Carlos Camacho who have been in charge of editing the text and providing assistance gathering the information cited in this study. As always, any remaining errors are the author's sole responsibility.
4. *Scrub* refers to vegetation consisting of herbaceous, woody and occasionally invading arboreal species of 1–3 metres high. It is no higher than 5 metres and grows on deforested, abandoned land. A *thicket* is also made up of vegetation consisting of herbaceous, woody and arboreal species, which are no more that 15 metres high, growing on land which was used for crops or pasture and then abandoned. It generally includes pioneer species which grow rapidly and their economic potential depends on the species available. Both scrubland and thickets may contain scattered trees of different sizes which may be used. Again, their economic potential depends on the particular species.
5. Since this chapter was written, and following the introduction of Forestry Law No. 7575,

new Payments for Environmental Services are proving to be the most innovative economic instrument favouring Costa Rica's forest sector.

REFERENCES

Ballestero, C. Lesmes (1996), 'Memoria Estadística del Sector Energía de Costa Rica', Publicación de la Dirección Sectorial de Energía No. 162, Ministerio de Ambiente y Energía, Enero.

Cámara Costarricense Forestal (CCF) and Ministerio de Economía Industria y Comercio (MEIC) (1995), 'Diagnóstico del Sector Industrial de la Madera: "Aserraderos y Fabricación de muebles"', Departamento de Fomento y Modernización Industrial del MEIC y la CCF, unpublished document, San José, Costa Rica.

CCT and CINTERPEDS (1996), 'Valoración Económico Ecológica del Agua: primera aproximación para la nteriorización de los costos en Costa Rica', mimeo, Heredia, Costa Rica.

Carranza, M. *et al.* (1996), *Valoración de los Servicios Ambientales de los Bosques de Costa Rica*, Centro Científico Tropical.

Gottfried, Robert R. (1995), *Economics, Ecology and the Roots of Western Faith: Perspectives from the Garden*, Garnham, MD: Rouman & Gottfried.

Kishor, N. and L. Constantino (1993), 'Manejo Forestal y Competitindad de Usos de la Tierra: un análisis económico para Costa Rica', *Región Latinoamericana*, Banco Mundial, October.

Ortiz, Rosalba (1996), 'El abandono de pastizales y el surgimiento del bosque secundario en Costa Rica', MA thesis, Universidad Nacional, Heredia, Costa Rica.

Pasos, Rubén (1994), *El último despale ... La frontera agrícola Centroamericana*, Diriamba, Nicaragua.

Pedroni, Lucio (1992), *Diagnóstico Forestal de Costa Rica.*, Informe de IC Y UICN-ORCA para la Cooperación Suiza para el Desarrollo (COSUDE), 31 de Agosto de 1992 (sin publicar).

Segura, Olman (1996), *Políticas Forestales en Centroamerica: Análisis de las Restricciones para el desarrollo del Sector Forestal en Costa Rica.* Elaborado por CINTERPEDS (UNA) y CCBAP, Heredia, Costa Rica.

Segura Olman, Miriam Miranda, Robin Gottfried and Luis Gámez (1996), 'Políticas Forestales en Costa Rica', investigación preparada para el Consejo Centroamericano de Bosques y Areas Protegidas (CCAB-AP), Heredia, Costa Rica.

Segura, Olman (1997), 'Síntesis de las Políticas Forestales en Centroamerica: Restricciones para el Desarrollo del Sector', O. Segura, D. Kaimowitz and J. Rodríguez (eds), San Salvador, El Salvador.

Solórzano, R., R. De Camino, R. Woodward, J. Tosi, V. Watson, A. Vasquez, G. Villalobos, J. Jiménez, R. Repetto and W. Cruz (1991), 'Accounts Overdue: Natural Resource Depreciation in Costa Rica', World Resources Institute, Centro Científico Tropical.

Stewart, R. (1992), 'An Economic Study of Costa Rica's Forest Sector', San José, Costa Rica.

UICN/IUCN (1996), *Diagnóstico Forestal de Costa Rica*, International Union for Conservation of Nature, Regional Office for Central America.

APPENDIX: COSTA RICA: AREA WHICH HAS BEEN REFORESTED, MANAGED AND PROTECTED WITH FOREST INCENTIVES (1972–95) IN HECTARES

Year	Income tax	CAF	CAFA	CAFMA	FDF	Trust 04–87	19–91	Trust 19–91 addendum	178 **	CPB	Article 63	Article 87	Total
1972	0.0	0.0	0.0	0.0	0.0	0.0	0.0	0.0	0.0	0.0	12 173.1	0.0	12 173.1
1978													0.0
1979	632.6	0.0	0.0	0.0	0.0	0.0	0.0	0.0	0.0	0.0	2 697.9	0.0	3 330.5
1980	1 073.8	0.0	0.0	0.0	0.0	0.0	0.0	0.0	0.0	0.0	294.1	0.0	1 367.9
1981	1 402.4	0.0	0.0	0.0	0.0	0.0	0.0	0.0	0.0	0.0	405.1	0.0	1 807.4
1982	877.3	0.0	0.0	0.0	0.0	0.0	0.0	0.0	**	0.0	0.0	0.0	877.3
1983	1 747.9	0.0	0.0	0.0	0.0	0.0	0.0	0.0	**	0.0	138.0	0.0	1 885.9
1984	1 194.1	0.0	0.0	0.0	0.0	0.0	0.0	0.0	**	0.0	4 520.8	0.0	5 714.9
1985	1 479.0	0.0	0.0	0.0	0.0	0.0	0.0	0.0	1 548**	0.0	69.1	0.0	3 096.0
1986	3 796.1	0.0	0.0	0.0	0.0	0.0	0.0	0.0		0.0	677.1	0.0	4 473.2
1987	4 753.8	0.0	0.0	0.0	0.0	0.0	0.0	0.0		0.0	1 102.3	286.0	6 142.1
1988	7 261.4	28.9	761.0	0.0	0.0	0.0	0.0	0.0		0.0	221.9	0.0	8 273.2
1989	5 668.1	943.3	1 269.0	0.0	1 876.0	0.0	0.0	0.0		0.0	760.3	1 001.0	11 517.7
1990	4 509.9	3 179.9	2 375.0	0.0	1 274.5	259.0	0.0	0.0		0.0	353.0	2 253.6	14 205.0
1991	977.9	4 616.2	3 217.0	0.0	2 366.0	120.0	0.0	0.0		0.0	1 513.0	4 486.6	17 296.6
1992	145.2	5 602.7	4 230.0	74.0	2 767.5	98.0	0.0	0.0		0.0	440.2	2 398.3	15 755.9
1993	0.0	5 788.0	4 135.0	75.0	2 767.5	0.0	0.0	0.0		0.0	1 569.9	3 390.4	17 725.8
1994	78.0	6 413.0	5 332.2	9 971.4	1 427.5	0.0	46.1	155.0		0.0	3 469.6	1 176.0	28 068.8
1995*	0.0	11 515.0	1 2499.1	12 000**	310.8	70.0	74.0	431.9		22 199.8	742.1	1 080.2	60 922.9
Total	35 597.4	38 087.0	33 818.3	22 120.4	12 789.8	547.0	120.1	586.9	1 548.0	22 199.8	31 147.2	16 072.1	214 634.0

Notes:
* In 1995 for CAF and CAFA, these hectares were approved by decree.
** Between 1982 and 1985, 1548 ha were planted through Trust 178.
CPB: Forest Protection Certificate.
CAFA: Advance Forest Bond Certificate.
CAFMA: Forest Mangaement Bond Certificate.
FDF: Forest Development Fund.
Art. 63: Area inscribed in the Forest Regime for Protection Purposes, Forest Law 7174.
Art. 87: Individual Resources

Source: National System of Conservation Areas, MINAE, 1996; FONOFIFO-MINAE, 1996.

10. Forest management in a semi-arid region in Brazil: a case study of the state of Rio Grande do Norte

Sebastião Kengen, Frans Pareyn, Newton D.E. Barcellos and Francisco C.B. Campello

INTRODUCTION

Environmental matters have received unprecedented attention in recent decades. Within this context, forest resources, especially their conservation and management, have deserved attention by scientists and policy makers. The primary focus has been on tropical rainforests. For instance, in Brazil the focus has been concentrated on the Amazon rainforest.

It is important, however, to note that Brazil encompasses a great diversity of climate and soils and, consequently, its forest resources are very diverse too. They range from tropical rainforest to subtropical forests. In spite of their environmental and socioeconomic importance, evidence suggests that most of them have not received the attention that they deserve. In short, it should be emphasized that they are as important and as fragile as the Amazon rainforest.

Forestry interventions have traditionally viewed the biomass problem as a simple supply and demand problem. Such thinking suggests that forest resources have been depleted due to a lack of balance between demand and supply. What can be generally noted is that extraction is much greater than what the environment can produce in a sustainable manner. From this perspective, the obvious solution is to plant as many trees as possible. Thus establishment of forest plantations, generally on a large scale and as monoculture crops, has been advocated.

In Brazil, this approach found a favourable environment in its model of development that had as a major aim to transform the Brazilian economy from an agricultural economy to a more industrial one (Kengen, 1985). Within this context, large-scale industrial forest plantations were established all over Brazil. It is estimated that about 6 million hectares were planted. Approximately 60 per cent of them were concentrated in the southeast and south regions. These plantations were established regardless of the environmental and socioeconomic characteristics of the region where they were established. Moreover, the impact of these plantations has not been adequately assessed.

In the Brazilian semi-arid region known as the 'Drought Polygon', it is estimated that approximately 350 000ha of forest plantations have been established. However, from this total only about 15 per cent can be considered to be in good condition (UNDP/FAO/IBAMA/BRA–87/007). This failure suggests that forest plantations *per se* are not the answer to forest problems in this region. Thus, if forestry development is to be promoted, it is very important to deal with it within a holistic perspective. In this context, forest management of the native vegetation seems to be one of the approaches to do so.

This chapter presents an overview of the results and recommendations made by different studies that were carried out among these projects. The emphasis will be on the state of Rio Grande do Norte, where the first efforts to analyse and develop forestry management techniques in the 'Caatinga'[1] were undertaken.

CONTEXT

A Brief Introduction to the Brazilian Northeast Region

The Brazilian Northeast Region is one of the five great regions of Brazil. It lies between 01° and 18° latitude south and 35° and 48° longitude west. It encompasses a total of nine states, namely Alagoas, Bahia, Ceará, Maranhão, Paraíba, Pernambuco, Piauí, Rio Grande do Norte and Sergipe. It covers an area of 1 561 178 square kilometres which corresponds to 18.27 per cent of the total area of Brazil. About 50 per cent of its total is located within the semi-arid 'Drought Polygon'. Its population is estimated to be 42 497 540 inhabitants, of which about 60 per cent live in urban areas and 40 per cent in rural areas (FIBGE, 1994). Socioeconomic factors as well as periodical droughts have contributed to a strong migration of its population, especially the rural one, to other parts of the country.

As far as economic development is concerned, the northeast is a mosaic of areas as heterogeneous as Brazil in which development has not evolved equally, nor have the benefits of this development been distributed evenly. It has been considered one of the poorest regions in Brazil. This situation has led this region to become the focus of attention of the federal government that has designed and implemented plans, programmes and projects that can contribute to promoting its economic development.

The northeast weather is unpredictable. The rain changes each year, in terms of quantity and distribution. Droughts are very frequent; for example, 33 occurrences have been registered since as early as the 18th century, of which 10 have been considered extreme. The natural vegetation is closely linked to the pattern of precipitation. Thus, as the precipitation varies through the region, the vegetation also varies, from sub-humid tropical forest, residues of which can still be found in the northeast coastal region, to 'cerrado' (savanna) that is common to central Brazil, to a thornbush vegetation called 'caatinga' that dominates the landscape of the 'Sertão'.

The importance of the rain is such that it helps to determine the four main zones into which the northeast region can be divided, as follows.

The first is the coastal zone, also known as the forest zone (*zona da mata*) which is a narrow, humid belt (100–200km wide) along the coast from the state of Rio Grande do Norte up to the south of the state of Bahia. In this zone one can still find some vestiges of the tropical rainforest, which is known as 'Mata Atlântica' (Atlantic Forest), that used to cover it. The rainfall is evenly distributed and it can reach up to an average of 2000mm per year. Major economic activities are concentrated in this zone.

The second zone is the 'Agreste', a transitional stage between the humid 'litoral' and the hot and dry 'Sertão'. The third zone is the 'Sertão' that is a great area inland with hot and dry weather subject to long periods of dryness and normally experiencing severe water shortages. It represents nearly 50 per cent of the total area of the northeast and its population is about 48 per cent of the total population. The major part of this zone has an average annual rainfall from 400 to 600mm, but it can be only 300mm in some areas. This rainfall is unevenly distributed through the year and it is concentrated in a period of three to four months. Agricultural activities are mainly restricted to subsistence due to periodical drought, among other factors.

The fourth (middle-north) zone is of high precipitation and covers parts of the states of Maranhão and Piauí. It is also a zone of transition between the semi-arid area and the Amazon basin. Precipitation here is between 600 and 2000mm.

The natural vegetation is closely linked to the pattern of precipitation. Thus, in the eastern coast of the northeast, one may still find residues of sub-humid tropical forest that is know as 'Mata Atlântica', as well as 'cerrado' (savanna), common to central Brazil. In the middle-north one also finds the 'cerrado' and palm forests. In the 'Sertão' a thornbush vegetation called 'caatinga' dominates the landscape.

Land Tenure and Use

Land tenure in the northeast has an asymmetrical distribution and has a concentrated structure, as can be noted from Table 10.1. However, it should be stressed that these characteristics are not specific to this region. Indeed, the land tenure in Brazil as a whole follows a similar pattern. This can be considered an 'Achilles' heel', since the quality of life in the region is directly related to the existing land ownership structure, that is to the concentration of the rural and urban estates and, consequently, to a highly unequal income distribution.

Table 10.1 Number and area of farms according to size

Area (ha)	Number	%	Area (10^3ha)	%
0–10	1 182 072	70	4 777.7	5
10–100	630 489	24	20 086.6	23
100–1,000	139 863	5	34 956.8	40
> 1,000	10 081	1	27 405.9	32
Total	2 592 505	100	87 277.0	100

Source: FIBGE (1994).

Table 10.2 Land use according to farm size

Area (ha)	Agriculture		Pasture		Forest		Not in use	
	Area	%	Area	%	Area	%	Area	%
0–10	3 379.7	71	728.9	15	245.2	5	218.4	5
10–100	6 127.4	31	6 805.9	34	3 371.5	17	2 781.8	14
100–1,000	6 056.9	17	14 492.1	41	7 714.0	22	5 482.4	16
> 1,000	2 446.7	9	11 562.0	42	7 439.0	27	4 329.0	16
Total	18 010.7	21	35 588.9	39	18 769.7	22	12 811.6	

Source: FIBGE (1994).

In Table 10.2, one can note that the use of the land for agriculture decreases while farm size increases, whereas for pasture and forest the opposite is the case. One possible explanation for this phenomenon is land speculation.

'Caatinga': a Brief Description

The 'caatinga' is a thornbush vegetation that occurs in a semi-arid area of the Brazilian northeast and it is basically composed of a woody shrub vegetation. The 'caatinga' does not have the exuberance of the rainforest of the Amazon basin. However, this does not mean that it has less value either in socioeconomic or in ecological terms. It is also just as fragile as the forests of the Amazon basin. Thus its importance should be understood within the context in which it occurs. Its exploitation and conservation therefore deserve the same attention that is given to the tropical rainforest.

However, similarly to what happens to a rainforest, the 'caatinga' has been depleted along the years and it is under great pressure. This fact can be mainly attributed to socioeconomic conditions of the region. In other words, its depletion can be linked to problems of the provision of basic needs. Thus its depletion cannot be perceived just as an ecological problem but it is also a socioeconomic problem, and consequently its conservation requires measures that can also contribute to alleviating poverty.

The major uses of the 'caatinga' are wood production; fuelwood, charcoal, posts and poles and wood for small rural constructions; non-wood forest products (NWFPs) and services; fodder, honey, fruits and medicinal plants. Its major importance is due to the overall utilization of the 'caatinga' as a natural pasture for extensive livestock and the energy supply for industrial and domestic use. About 70 per cent of the families use firewood and charcoal as sources of energy. Indeed, firewood represents from 23 per cent to 41 per cent of the energy source. Firewood is the second most important source of energy in the region, the first being oil.

Forestry Potential of the 'Caatinga'

The native vegetation in the northeast covers about 45 per cent of the total area of the northeast. The 'caatinga' covers about 60 per cent of the area of the 'Sertão'. This suggests the great importance of the 'caatinga'. Allied to this, given the edaphoclimatic conditions of the 'Sertão', evidence suggests a low performance of forest plantations of exotic species. In this context, the management of the native vegetation represents a great potential, can have positive environmental impacts and can contribute to reducing deforestation.

MANAGEMENT OF THE 'CAATINGA'

Specific Characteristics of the 'Caatinga' Management

Conclusions from research since 1982 were that the natural vegetation was in an accelerated process of degradation. This necessitated the making of efforts to avoid, or at least diminish, this process of degradation. One of the solutions proposed was the development and introduction of management techniques to promote sustainable use of the 'caatinga'. However, given the socioeconomic conditions of the 'Sertão', the management of the 'caatinga' should involve low-cost forest management techniques. It is important to point out that the development and introduction of such techniques constituted a great challenge owing to the lack of tradition and know-how with regard to managing the 'caatinga'. Thus it was not just a matter of applying or introducing forest management techniques: it was necessary to develop, adapt and disseminate techniques that could address these conditions.

Within this context, the following were proposed in 1987:

- techniques of exploration: clearcutting;
- theoretical basis: forestry inventory and analysis by diameter class and species;
- cutting cycle (rotation): 15–20 years;
- medium annual increment (MAI): an adequate estimate does not exist, but, based on the cutting cycle and total stock, an average of seven steres/ha/year is a good approximation;
- silvicutural treatments: the use of fire is not recommended as its use can cause damage that retards natural regeneration; the use of the area managed as pasture is suggested, but the number of cattle per hectare should be controlled;
- objective: mainly firewood production with, on a smaller scale, poles and other construction materials.

It is important to stress the need for integrating the management of the 'caatinga' into an agroforestry system in order to encourage farmers

Present Management Technology

The natural regeneration of the 'caatinga' is not yet very well understood. The available knowledge is restricted to empirical observations. The major regeneration forms are (a) via seeds, (b) via sprouts of the stumps, and (c) via sprouts of the root. It should be noted that the possibilities of regeneration of the 'caatinga' are good despite the rough climatic conditions.

Clearcutting is one of the techniques used and it can be followed by fire, without fire and by leaving some trees for future seed collection, which is an improvement over the clearcutting technique. Selective cutting can depend on the diameter of the tree and on its species. It should be accompanied by leaving some trees for future seed collection.

It is important to note that, although these techniques have been suggested and applied, the management of the 'caatinga' still requires more research in order to understand its full potential.

Socioeconomic and Ecological Benefits

The major socioeconomic factors associated with the 'caatinga' can be summarized as follows:

- low cost,
- production of wood and NWFPs,
- labour can be used on the 'caatinga' during drought periods to reduce rural migration,
- it contributes to improved soil conservation,
- it contributes to biodiversity conservation.

MANAGEMENT OF THE 'CAATINGA': AN EXAMPLE

The use of wood as a source of energy is very important both at domestic and industrial level. It can be used as firewood in boilers. This consumes about 24 000 steres per year of firewood, which corresponds to a demand of two or three trucks of 20 steres each per day. This firewood can be bought in the market, though this may be difficult during the rainy season, as most of the population is engaged in agricultural activities. The forestry inventory in volume is estimated at about 115 to 129st/ha.

On the basis of the inventory, the final use and local conditions, the company decided that the management should take into account the following criteria:

- exploration using selective cutting of trees above 6cm in diameter,
- cutting of all non-fodder species,
- conservation of all species with potential to go to the sawmill,
- rotation of 10 years,
- production per hectare of about 97 steres,
- regeneration via seeds and sprouts of the stumps.

Economic aspects of the management in this example can be summarized as follows:

- price of the firewood in the market at the plant door (average): US$6 /st,
- value of the stand firewood (average): US$2st,
- cost to cut the firewood (average): US$1/st,
- freight (from farm to plant): US$3/st.

While the firewood bought in the market costs about US$6/st, the firewood produced by the management of native vegetation costs about US$4/st (cut plus freight).

FINAL REMARKS

We can summarize the results as follows. First, the management of the 'caatinga' is still in its initial phase and still requires more research. However, evidence suggests that it can be viable and it seems to be better than forest plantations in such conditions. Second, its utilization has important socioeconomic and environmental implications that deserve atttention. Third, the management of the 'caatinga' showed such potential that the 'Banco do Nordeste' decided to start a credit line to make it possible for farmers and companies to promote forest management in the 'caatinga'. This is a great advance when one considers (a) that traditionally support has been for forest plantations, although evidence suggests that most of them have failed owing to the harsh edaphoclimatic conditions of the northeast, particularly in the 'Sertão', and (b) that there was no incentive, least of all financial, for management of the 'caatinga'.

NOTE

1. 'Caatinga' is a word from the 'tupi-guarani' language spoken by the Brazilian Indians and it means white or grey forest. According to Riley and Young (1966) 'caatinga' is a plant community (thornbush) dominated by thorny trees or shrubs called thorn scrub. It is a xerophytic vegetation typical of dry environments. Density (stems/ha) depends on factors such as rainfall and soil.

REFERENCES

Fundação Instituto Brasileiro de Geografia e Estatística (FIBGE) (1994), *Anuário Estatístico do Brasil – 1994*. v. 54, Rio de Janeiro: FIBGE.

Kengen, S. (1985), 'Industrial Forestry and Brazilian Development: A social, economic and political analysis, with special emphasis on the fiscal incentives scheme and the Jequitinhonha Valley in Minas Gerais', unpublished PhD thesis, Australian National University.

Riley, D. and A. Young (1966), *World Vegetation*, New York and Cambridge: Cambridge Universtiy Press.

UNDP/FAO/IBAMA/BRA, O Setor Florestal do Nordeste e as Atividades do Projeto PNUD/FAO/IBAMA/BRA–87/007, unpublished report (mimeo).

Name index

Adger, N. 108, 117
Adinarayana, J. 43
Alder, D. 31
Alemdag, I.S. 66, 67
Andersen, L.E. 108
Andrasko 37
Andrasko, K. 107

Ballestero, C. 235
Barbier, E.B. 167, 171
Barnett, H.J. 79
Barres, H. 136
Barton, D.N. 166
Botkin, D.B. 63, 73
Bouman, O.T. 217
Brand, D.G. 203, 210
Brody, A. 84
Brown, G.M. 79, 83
Brown, S. 3, 5, 6, 37, 66, 107, 108, 114, 117, 126, 136, 184

Callaghan, B. 63, 66
Cámara Costarricense Forestal 245, 253
Carpenter, R.A. 217
Carranza, C. 108, 141, 191
Carranza, M. 232
CATIE/IUCN 178
CCFM 201
CCT and CINTERPEDS 232
Centro Cientifico Tropical 46, 48
Chiang, A.C. 101
Chinchilla, M. 184
Cline, W.R. 141
Collinet, J. 170
CONAP 183
Constantino, L. 232
Corporación de Fomento Ganadero (CORFOGA) 117
Costa Rican Office for Joint Implementation (OCIC) 110, 112
Costanza, R. 188, 189
Current, D. 40

Dale, V.H. 156
David, M.B. 66
D'Croz, L. 178
De Groot, R. 175, 182
Detwiler, R.P. 156
Dixon, J.A. 107, 173, 177
Dixon, R.K. 63, 156, 161, 185
Doell, S.W.M. 205
Dore, M.H.I. 81, 83
Dugan, P.J. 177

Eckman, K. 217
Environment Canada 156

Faeth, P. 136
Fan, S. 63
Fankhauser, S. 110, 141
FAO 2, 80
Farzin, Y.H. 80, 84
Fassbender, H. 6
Faustino, J.M. 42
Fearnside, P. 108
FIBGE 262, 264
Field, B.C. 79, 83
Finnegan, B. 126
Fischlin, A. 63
Fisher, A.C. 79
Freedman, B. 156
Freese, F. 66
Furst, E. 166

Galvez, J. 169, 170, 183, 184
Godoy, J. 4
Goldberg, A.S. 116, 121, 123
Gómez, M. 120
Gottfried, R.R. 258
Grace, J. 3
Grigal, D.F. 66, 70, 72

Hall, C.A.S. 156
Hall, D.C. 79
Hall, J.V. 79
Halliwell, D.H. 63
Hanson, D.A. 79
Harmon, M.E. 71
Haugen, C. 4
Heal, G.M. 79
Heinselman, M.L. 72
Heinzman, R. 183
Horbulyk, T.M. 103
Houghton, R. 4, 63, 108, 117, 156
Hutchinson, I.D. 33

IICA/FLACSO 169
IMFN 203, 221
IPCC 4

Jimenez, F.O. 42
Johnston, M.H. 66, 72, 79, 161
Judge, G.G. 118

Kaimowitz, D. 117
Kalra, Y.P. 66
Kauppi, P. 3
Keith, T. 156
Kengen, S. 261
Kirschbaum, M. 63
Kishor, N. 232
Kooten, G.C. van 103, 161
Krinsky, I. 115, 121
Kulshreshtha, S.N. 159, 205, 206, 207, 208, 221
Kurz, W.A. 63, 73, 82, 104, 211
Kwiecinski, B. 178

Lafranchi, S. 27
Leonard, D. 101
Loewen, K.G. 208
Lugo, A. 3, 5, 6, 37, 66, 136, 169, 184
Lutz, E. 40, 181
Lynne, G. 188

McNeely, J.A. 177
Maguire, D.J. 42
Malheiros, W. 108
Maltby, E. 177
Marland, G. 85, 86, 91, 163
Marmillod, D. 169, 170
Martinez, H. 6
Maynard, D.G. 66
Medina, J.M. 45
Merriam, H.G. 63
Ministerio de Recursos Naturales, Energia y Minas 138
Ministerio del Ambiente y Energia (MINAE) 110, 137, 245, 247, 248
Mohren, G.M. 136, 163
Moriarty, B.F. 206
Morse, C. 79
Munasinghe, M. 181

Nabuurs, G.J. 136, 163
Nalder, I.A. 63
Natural Resource Canada 201, 203, 221
Nault, R. 202
Nilsson, S. 4
Nordhaus, W.D. 79

Ohmann, L.F. 66, 72
Ortiz, R. 37, 231

PAMF 204
Parfit, D. 157
Pasos, R. 225
Pastor, J. 63
Pearce, D. 166
Pearce, D.W. 177
Pedroni, L. 231
Picado, V. 6
Post, W.M. 63
Prado, M. 169, 178, 180

Quirk, J.P. 157
Quiros, D. 126

Ramanathan, R. 105
Ramirez, O.A. 116, 120
Reining, C. 183
Reyes, R. 171
Richards, F.J. 114
Robb, A.L. 115, 121
Rojas, E. 169
Rolston, H. 222
Rosenthal, D. 179
Ross, W.A. 103

Saenger, P. 178, 180, 189
Salazar, R. 6
Sanford, R. 182

Saposnik, R. 157
Saskatchewan Agriculture and Food 158
Schlamadinger, B. 63
Schlesinger, W.H. 161
Schophauser, W. 4
Schroeder, P.E. 107, 136
Sedjo, R.A. 3, 156
Segura, M. 37
Segura, O. 225, 226, 227, 229, 230, 231, 245, 249
Sen, A. 157
Sención, G. 175, 176, 181, 182, 183, 184, 185
SERM 150
Sherman, P.B. 177
Shields, J.A. 151
Shultz, S. 54
Siemens, J.K. 205
Silva, C. 169, 178, 180
Simpson, L.G. 63, 73
Sims, R.A. 63, 64, 70, 71, 72
Slade, M.E. 79
Smith, V.K. 79
Snedaker, S.C. 169, 189
Solórzano, R. 117, 232
Solow, R.M. 79
Squire, L. 186
Stadmüller, T. 22, 24, 25
Statistics Canada 158
Stewart, R. 244
Swaine, M.E. 30, 31

Tak, H. van der 186
Tans, P.P. 63
Theocharoppolos, S.P. 43
Thoranisorn, S. 136
Tol, R. 110, 141
Torres, G. 135, 136
Trexler, M. 4
Turcios, W. 24, 25
Turner, D.P. 66
Turner, R.K. 177

Uhlig, P. 64
UICN/IUCN 243, 244
UNDP/FAO/IBAMA/BRA 262

Vitousek, P. 182

Walker, H.V. 205, 206
Wan, F.Y. 79
Ward, A.J. 81
WCED 202
Wedeles, C. 211
Weitzman, M. 157
Weyerhaeuser Canada Ltd 222
Wiken, E.B. 150
Windevoxhel, N.J. 169, 175, 176, 177, 178, 179, 180, 188, 189
Winjum, J.K. 141

Young, A. 6

Subject index

Aboriginal Employment Equity Program 210
aboriginal families, wilderness preservation 208
Accounts Overdue: Natural Resource Depreciation in Costa Rica 232
advanced forest bond certificates 250
afforestation 7
Agreste 263
agriculture, versus forestry study 148–64
agroforestry
  carbon storage 5, 6
  costs associated with 52
air photography 56
ARIMA model 94–100
aspen 72, 162, 206
attributes, mangrove forests 187
Avicennia spp. 169

balsam fir 67, 72
Banco do Nordeste 268
bark resources, mangroves 180–81
basal area distribution, silvicultural interventions 21, 23
biodiversity
  attributes, mangrove forests 187
  conservation 20–21, 22
biomass accumulation
  primary forests, Costa Rica 126–7, 128
  secondary forests, Costa Rica 114–17, 118–21
biophysical data, integrating socioeconomic and 41–4, 57–60
biophysical processes, resource management 40
black ash 71
black spruce 67, 70, 206
boreal forests
  carbon sink model
    assumptions 81
    conclusions 100–101
    econometric estimates 85–100
    implication of assumptions 82
    modelling carbon sequestration 82–3
    modelling shadow price 83–5
  versus agriculture, Saskatchewan 148–64
Boreal Plains, Saskatchewan 149–51, 160–61
boundaries, georeferencing 56–7
Brazil, forest management 261–8
  Caatinga 265–8
  land tenure and use 264
  northeast region 262–3
  remarks on 268

Caatinga
  benefits 267
  description 265
  management
    characteristics 266
    example 267–8
    technology 266–7
  potential 265
cadastral based maps 44, 56
Canada
  boreal forests, carbon sink study 79–106
  see also Ontario; Saskatchewan
Canadian Model Forest Program 201
  expansion to international model 203
  origin and establishment of 202–3
  selected characteristics 224
  see also Prince Albert Model Forest
canola 155
carbon budget, human perturbations 4
carbon cycle
  boreal forests as carbon sink 79–106
  tropical forests as carbon sink 107–45
carbon dioxide, global warming 107

carbon emissions
anthropogenic (1995) 3
ARIMA model 94–5
Canadian (1961–91) 85–6
control costs 141
deforestation 108
social costs 109, 110
trend analysis 86–8
unit root tests 105
carbon management, options 5, 6, 7
carbon parking 109
compensation, cross-country 144
payments 110
values, secondary forests 124
carbon pools
in ecosystems 3–4
sustainable management 4
carbon sequestration 109
Canadian forests 80
carbon management options 7
conversion of agriculture to forest 161
cross-country compensation 145
land availability 4
mangrove forests 184–5
modelling 82–3
rate of 103–5
secondary forests 124, 125, 142, 144
carbon sinks, forests as 37
boreal forest case study 79–106
tropical forest case study 107–45
carbon storage 109
agroforestry, Central America 5, 6
payments 110
plantation forestry, Costa Rica 143–4
primary forests, Costa Rica 142
secondary forests, Costa Rica 10, 142
vegetation and soil, Ontario 63–73
casuarina 50
cedar 50
Cedrela odorata 167, 183, 191
Central America
agroforestry, carbon storage 6
sustainable resource management 40–62
valuation of mangrove forests 166–95
see also Costa Rica
Certificates of Tradable Carbon Offsets (CTOs) 110–11, 112
Chamadorea spp. 183, 191
clearcutting 267
climate change, land use 156
Cochrane–Orcutt model 89, 91
commercial forestry, Saskatchewan 155–6, 158
commodity prices, Saskatchewan 154–5
compensation, Costa Rica 124, 125, 143, 144, 145
competitiveness, timber industry, Costa Rica 258
construction industry, Costa Rica 227, 246
conventional harvesting (control) 10, 128–34, 142–3
Cordia alliadora 136
correct use areas 47
Costa Rica
economic development 225
forest resources
differing uses 227
failure to recognize importance 227–30
in national economy 231–3
forest sector 226–7
constraints on 253–4
potential development 254–8
joint implementation agreement 110, 111–13
payments, sustainable management 143
plantation forests 231
primary forests 230
reforestation
areas 260
incentives 249–52
Purire River watershed 44–52
resource planning, Pacuare watershed 52–60
silviculture 233–5
social actors and institutions 252–3
timber industry 235–49
tropical forests, carbon sink study 107–45
Villa Mills Demonstrative Experimental Area 19–38
costs, Purire River watershed, reforestation 51–2
critical land use areas, identification with GIS 40, 42–3
Pacuare watershed 54–6
Purire River watershed 45–9

crops, Saskatchewan 154, 155
cross-country compensation 144, 145
Crown lands, PAMF 209
Crowsnest Pass Agreement Rates 148
cultivation, Saskatchewan 153–4
Cyathea gracilis 20, 29
cypress 50

data entry, GIS 41
data manipulations, GIS 42
Decision Support System (DSS) programme 211–12
deforestation
  carbon emissions 108
  carbon sequestration 7
  Costa Rica 134, 230
  tropical 2
degradation cost 79
degradation data, Pacuare Watershed 55–6, 60
degraded lands, Pacuare watershed 54–5
Dendropanax querceti 21
depreciation, forest destruction, Costa Rica 232
Desmoncus sp. 183–4, 191
Diagnóstico Forestal de Costa Rica 244, 246
Dickey–Fuller unit root test 91, 92
Drought Polygon 262

ecological benefits, Caatinga 267
ecological functions, forests 1
ecological services, valuation of 173
economic activities, PAMF 206–8
economic globalization, forest industry 249
economic valuations
  agriculture versus forestry 148–64
  boreal forests as carbon sink 79–106
  mangrove forests 166–95
  tropical forests as carbon sink 107–45
ecosystems
  carbon pools 3–4
  research 209
  valuation of 171–3
ecotourism 27
El Buen Precio S.A. 247
employment income generation 25–6
environmental conditions, carbon storage 70–71
environmental services
  economic value 108
  forest functions 37
  forest sector, Costa Rica 227, 232, 254
  mangrove forests, goods 194
  non-marketed 1
erosion control, mangrove forests 185–6, 192–3
Espaveles demonstrative site 113, 114, 122
Eucalyptus spp. 50, 135, 136
export industries, timber products, Costa Rica 247
extractive activities, mangrove forests 169

farmers, preferences, reforestation 49–50
farms
  Brazil, northeast region 264
  Pacuare watershed
    georeferencing 56–7
    prioritization exercise 60
    soil conservation programme 57–60
  Saskatchewan 152
felling 26
Financial Products Center Ltd 112
financial resources, PAMF 208
fires, carbon storage 72
First Nations people 209, 210, 216
fiscal incentives, reforestation 250–51
fishing, mangrove forests 179–80, 181, 188
flooding, Purire River 44–5
flora and fauna, forest management 20–21, 22
Florencia demonstrative site 114, 119, 122
forest bond certificate 250
forest conservation certificate 249–50, 251
Forest Ecosystem Classification (FEC) 63, 64, 65, 66, 67, 68, 72
forest management
  Brazil, case study 261–8
  forest functions 19–38
  group consensus approach 217
  partnership approach 216–17
forest management bond certificate 250

Forest Management License Agreements (FMLA) 155, 209
forest policy, Costa Rica 225–58
forest protection certificate 250–51
Forest Resource Inventory (FRI) 66–7, 72
forest sector, Costa Rica 226–33
  constraints on 253–4
  potential for development 254–8
forestland, use of 6
forests
  ecological functions 1
  functions and forest management 19–38
  need for analysis of 1–2
  source of GHGs 3, 5
  see also agroforestry; boreal forests; mangrove forests; tropical forests
Framework Convention on Climate Change (UN) 1, 2, 3, 108, 161
fuelwood 26, 27, 28, 169, 180, 207, 235, 237

gameboards, Costa Rica 246
General Forest Administration 230, 245
geographic information systems (GIS)
  case studies
    conclusions 61–2
    reforestation 44–52
    resource planning 52–60
  identifying critical land use areas 42–3
  PAMF 209
  socioeconomic data, planning process 44
  technology 41–2
Global Forest Convention 1
global positioning systems (GPS) 44
global warming
  carbon concentrations 107
  recognition of problem 1
  reforestation 162–3
Gmelina arborea 135, 136
goods, mangrove forest
  Guatemala 182, 183–4, 191–2, 193
  Nicaragua 178–9, 189–90
grain farming, analysis, Saskatchewan 158
greenhouse gases (GHGs)
  forests as source of 3, 5
  stabilizing concentrations 3
gross domestic product (GDP), value of forests 89, 90, 93, 97
group consensus approach, forest management 217
Guatemala, mangrove forests 169–71
  economic problems 167
  valuation methods 181–7
  valuation results 191–3

harvesting, carbon storage 67, 70, 71
  conventional (control) 10, 128–34, 142–3
Héroes Mártires de Veracruz 168
Hieronyma alchorneoides 135
Hotelling model 79
human perturbations, carbon budget 4

Ian Hutchinson experimental site 113, 114, 119, 122
Ilex sp. 20
impact valuation 171, 172
incentives, reforestation 249–52
income tax deduction, reforestation 249, 250
indirect valuation 172–3, 193
industrialized countries
  cross-country compensation 144–5
  emission control costs 141
institutional processes, resource management 40
institutions, Costa Rica 252–3
integrated resource management, PAMF
  group consensus approach 217
  knowledge base 214
  legislative change 218
  limited success of 212–13
Intergovernmental Panel on Climate Change (IPCC) 3, 108, 144
International Institute for Environment and Development (IIED) 167
international model forests 203, 224

jack pine 71, 206
joint implementation programmes 10, 108, 110–13
jual 50

knowledge base, PAMF resource management 209, 214

Kyoto Principle 1, 161

labour market, Costa Rican timber industry 240–43
Laguncularia racemosa 169
land, availability, carbon sequestration 4
land tenure, Brazil, northeast region 264
land use
  Brazil, northeast region 264
  capacity maps 43, 46
  PAMF region 204–6, 209
  Saskatchewan (1921–76) 154, 156–63
  see also critical land use areas
legislative change, PAMF 218
liberation treatment 10, 128–34, 142, 143
logging 26, 207

mangrove forests, valuation 166–95
  conclusions 193–5
  methodology 171–87, 194
  results 187–93
  study areas 168–71
Manilkara spp. 170, 183, 191
Mann Whitney test 29
maple 71, 72
marginal social damage costs, carbon emissions 109, 110
marginal social opportunity costs
  ARIMA model 94–100
  conclusions 101
  trend analysis 89–94
market goods, PAMF 206
Mata Atlântica 263
match factories, Costa Rica 246
medicinal products, Costa Rica 227
Mid-Boreal Lowland (MBL) 150, 151
Mid-Boreal Upland (MBU) 150, 151
Ministerio de Recursos Naturales, Energía y Minas 138
monopsony 244
monotonic shadow price 86–9
Montreal Lake Cree Nation (MLCN) 206, 210, 213
mortality, silvicultural interventions 29, 31

National Council of Protected Areas (CONAP) 170
National Forest Yield Simulation Programme 126
National Investment Fund for Forest Financing 111
National System of Conservation Areas (SINAC) 230, 231, 249, 252
natural disasters 1
natural disturbances, carbon storage 72, 104
natural regeneration
  Caatinga 266
  Purire River watershed 52
  silvicultural interventions 24, 28–9
natural succession 161
Nicaragua, mangrove forests 168–9
  economic problems 167
  valuation methods 177–81
  valuation results 187–90
non-governmental organizations, Costa Rica 252–3, 256
non-market goods, PAMF 206, 208
non-timber products, Costa Rica 27, 226–7
nutrient cycling, mangrove forests 186–7, 192

oak 24, 27, 50
Ocotea sp. 20
Office for Joint Implementation (OCIC) 111
Olafo project 167, 169, 183
Ontario, carbon storage study
  data analysis 66–7
  discussion 70–73
  results 67, 68, 69, 70
  study area and sample design 63–6
ornamental plants, Costa Rica 227
overuse areas 46–7
ownership data, Pacuare Watershed 57

Pacuare watershed 52–4
  georeferencing farm locations 56–7
  identification, critical land use areas 54–6
  integrating data 57–60
PAMF *see* Prince Albert Model Forest
paper manufacturing, Costa Rica 238, 239, 240
partial valuation 171, 172
partnership approach, PAMF 204, 213

conflicts, decision-making 218–19
effectiveness of an extension 216
sustainable management 217
pasture, ecological impacts 160–61
payments
carbon storage or parking 110
Costa Rica
environmental services 232
sustainable management 143
pine 67, 71, 72, 206
plantation forestry
Costa Rica 231
carbon sink services 134–41, 143–4
Saskatchewan, carbon sequestration 161–2
plywood factories, Costa Rica 246
Podocarpus sp. 29
PORTICO 246–7
precautionary principle 1
presswood factories, Costa Rica 246
primary forests, Costa Rica 230
carbon sink services 126–34, 142, 143
Prince Albert Model Forest (PAMF)
accomplishments 210
analysis of 202
communication with public 220
conflicts, decision making 218–19
distinguishing features 209, 210
economic activities 206–8
endogenizing values 217–18
financial resources 208
formation of 203–4
group consensus approach 217
land use in 204–6
legislative changes 218
partnership approach 216
extension of 216–17
sustainable forest management 217
programme evaluation
effectiveness 213–14
impacts 215–16
issues 211
management 214–15
relevance 211–13
property rights 218
research 209, 213, 220
summary and conclusions 220–21
production data, Pacuare watershed 57
property rights, PAMF 218
PROSIBONA project 113–14
public
communication of PAMF results 220
perception of PAMF 214
pulp industry, Saskatchewan 156, 162
Purire River watershed 44–5
identification of critical land use areas 45–9
reforestation
estimating costs 51–2
incorporating farmers' preferences 49–50
proposed plan 51

Quercus copeyensis 20, 27–8, 29, 30, 34, 35, 36, 50
Quercus costaricensis 20, 27–8, 29, 30, 32, 33, 34, 35, 36, 50

rain interception, tree crowns 25
recreation
Costa Rica 27
mangrove forests, Nicaragua 179
PAMF 206
recruitment, silvicultural interventions 31
red maple 71
red pine 67, 71, 72
reforestation
carbon sequestration 7
Costa Rica 44–52, 249–52, 260
economic considerations 162–3
renewable resources, scarcity rent 79–80
research centres, Costa Rica 253
Research Committee, PAMF 214–15
research, PAMF 209, 213, 220
resource management
mangroves 180–81
planning with GIS 40–62
resource planning, Pacuare watershed 52–60
Rhizophora spp. 167, 169
Rio Grande do Norte, case study 261–8

San Miguel Community Concession 169–71
Saskatchewan, agriculture versus forestry 148–64
agricultural activity 152–5
Boreal Plains region 149–51
commercial forestry 155–6

land use
and global climate change 156
policy considerations 160–63
results 159–61
study and methodology 156–8
variables and sample period 158–9
soil capability 151–2
see also Prince Albert Model Forest
sawmill industry
Costa Rica 245–6
Saskatchewan 155–6, 162
scarcity rent 79–80, 83
secondary forests, Costa Rica 230–31
carbon sink services 113–26, 142, 144
carbon storage and sequestration 10
Sertão 263, 265, 266, 268
services, mangrove forests
Guatemala 182, 184–7, 192–3
Nicaragua 179–81, 188–9
shadow price, forests 80
conclusions 100–101
model results 98
modelling 83–5
trend analysis 86–9
shrimp fishing, mangrove forests 179–80, 181, 188
silviculture
biodiversity conservation 20–21, 22
Costa Rica 233–5, 236, 237
primary forests 126–34, 142–3
timber production 28–36
silvopastoral systems, carbon storage 5
social actors, Costa Rica 252–3
social benefits, land use 157, 158–60
social functions, forests 6, 25–6
socioeconomic benefits, Caatinga 267
socioeconomic data
in GIS planning process 44
identifying land use areas 40
integrating biophysical and 41–4, 57–60
socioeconomic processes, resource planning 40
socioeconomic research, PAMF 209
soft credit 251
soils
carbon in, tropical 6
carbon storage, Ontario 63–73
conservation, Pacuare Watershed 57–60
northern Saskatchewan 151–2
protection 22–4
spatial overlay 42, 43
spruce 67, 70, 72, 206
stochastic efficiency 157–8
subsidies, agriculture, Saskatchewan 160
sugar maple 72
Swietenia macrophyla 167, 183, 191

Talamanca Cordillera 24
technological development, Costa Rica 254
Technological Institute of Costa Rica 230
Tectona grandis 135–6
timber
Costa Rica 235–49
potential for development 256–8
production and sustainability 27–36
PAMF region 206, 207–8
Saskatchewan
commercial products 155
for pulp and lumber 162
valuation, mangrove forests 183–4, 191–2
Tirimbina demonstrative site 113, 114, 119, 122
toothpick factories, Costa Rica 246
Total Economic Value approach 206
total net social benefit, land use 157
total valuation 171, 172
tourism
Costa Rica 27
Nicaragua 179, 189
trade, forest products, Costa Rica 247–9
tree crowns
illumination and shape 33–5, 36
rain interception 25
tree diameter
crown illumination and shape 33–5, 36
silvicultural interventions 29, 30, 32
tree height, silvicultural interventions 29, 30
trees
by species, after silvicultural interventions 22
farmers' preferences 50
trend analysis
shadow price 86–9

social opportunity costs 89–94
Tropical Agricultural Research and Higher Education Center (CATIE) 45, 167
tropical forests
carbon sink model 107–45
estimating services 113–41
joint implementation 110–13
recommendations and conclusions 142–5
valuing services 109–10
carbon in soils 6
deforestation 2
Tropical Scientific Center 232

underuse areas 47
unit root tests 105–6
University of Maryland 167
upper canopy treatment 10, 128–34, 143

value added, Canadian manufacturing
ARIMA model 95
trend analysis 88
unit root tests 105–6
values
associated with PAMF 206
endogenizing, PAMF 217–18
forest per hectare
ARIMA model 96–100
trend analysis 89, 93, 94
vegetation, carbon storage, Ontario 63–73
Villa Mills Demonstrative Experimental Area 19–38
Virgen de Sarapiqui 126

Waskesieu 206
watersheds
conservation of water regime 24–5
see also Pacuare watershed; Purire River watershed
Weinmannia sp. 21
Weyerhaeuser Canada Ltd 155, 203, 209, 210
wheat, Saskatchewan 154, 155
white pine 67, 71, 72
white spruce 67, 72, 206
White's test 87, 88, 92
wilderness preservation, PAMF 208
wildlife species, Costa Rica 227
wood extraction, mangrove forests 180, 188–9
World Conservation Union (IUCN) 167

yellow birch 71